T0320943

# A Numerical Primer for the Chemical Engineer

## Second Edition

# A Numerical Primer for the Chemical Engineer

## Second Edition

Edwin Zondervan

CRC Press is an imprint of the
Taylor & Francis Group, an informa business

CRC Press
Taylor & Francis Group
6000 Broken Sound Parkway NW, Suite 300
Boca Raton, FL 33487-2742



Printed on acid-free paper

International Standard Book Number-13: 978-1-138-31538-9 (Hardback)



**Visit the Taylor & Francis Web site at**
**http://www.taylorandfrancis.com**

**and the CRC Press Web site at**
**http://www.crcpress.com**

Printed and bound in Great Britain by
TJ International Ltd, Padstow, Cornwall

*In memory of my dear father,*
*Ekke Zondervan.*

# Contents

# *Introduction*

Since 2008 I have been lecturing a course on numerical methods, at first at Eindhoven University in the Netherlands and since 2015 at Bremen University in Germany. The course is set up specifically for chemical engineers. There are good references for this subject, and everyone who has a heart for numerical methods probably possesses the book *Numerical Recipes* by William Press and co-authors.

When I started collecting material for the course, it began with the online material of John Hult (Cambridge), who allowed me to use his materials with the proper acknowledgments. I also discovered some materials concerning computer errors from Roel Verstappen (Groningen), which I thought could be important to incorporate in the course.

After development of MATLAB code, exercises, assignments, and lecture slides, I thought it could be handy to put everything into a syllabus, as a useful guide for the students attending the class.

In 2013, it turned into a real book and was published with CRC! After using the book successfully in the last few years, I noticed that several corrections and extensions were needed to further improve the book. For example, a more detailed discussion on model development in the first chapter was needed. A chapter on numerical integration was missing as well. I also thought a short description on boundary value problems could be useful, as I was discussing it in class, but it was not documented in the book.

After carefully making the modifications, I can proudly say that I am happy that a second edition of my book is now released. A numerical primer for the chemical engineer will give you a taste for this exciting field and prove that this discipline is alive and vivid. There is a reference list with useful books and papers. In case you are interested, you can contact me regarding MATLAB code, references, and other course materials.

I hope this second edition will be a useful companion. Enjoy!

# Preface

This book emphasizes the derivation and use of a variety of numerical methods for solving chemical engineering problems. The algorithms are used to solve linear equations, nonlinear equations, ordinary differential equations, and partial differential equations. It also includes chapters on linear and nonlinear regression, and on optimization. MATLAB®* is adopted as the programming environment throughout the book. MATLAB is a high-performance computing program. An introductory chapter on MATLAB basics has been added and Excel users can find a chapter on the implementation of numerical methods in Excel. Worked-out examples are given in the case study chapter to demonstrate the numerical techniques. Most of the examples were written in MATLAB and are compatible with the latest versions of MATLAB.

It is important to mention that the main purpose of this book is to give students a flavor for numerical methods and problem solving, rather than as an in-depth guide to numerical analysis. The chapters end with small exercises that students can use to familiarize themselves with the numerical methods.

The material in this book has been used in undergraduate and graduate courses in the chemical engineering department of Eindhoven University of Technology. To aid lecturers and students, course materials have also been made available on the Web at `http://www.crcpress.com/product/ISBN/9781138315389`.

Finally, the author would like to thank everyone who has been helpful and supportive in the creation of this book, especially some of the PhD students at Eindhoven University who have assisted during lectures and directly influenced the content of this book: Juan Pablo Gutierrez, Esayas Barega, and Arend Dubbelboer.

Edwin Zondervan
January 2014

*MATLAB is a registered trademark of The MathWorks, Inc. For product information, please contact:

The MathWorks, Inc.
3 Apple Hill Drive
Natick, MA 01760-2098 USA
Tel: 508 647 7000
Fax: 508-647-7001
E-mail: info@mathworks.com
Web: www.mathworks.com

# 1

# *The role of models in chemical engineering*

## 1.1 Introduction

The concept of a model has been around since ancient times. Models appear in all branches of science and engineering. However, it is often said that modeling is more art than science or engineering. In this chapter we will discuss general aspects of models, and more specifically the models that describe (chemical) process systems. It is not intended as an in-depth discussion.

Ultimately, this book is about solving the developed models in a numerical fashion. We could consider Ptolemy's *Amalgest* (150 BC) as one of the first recorded studies on modeling and numerical analysis in which numerical approximations to describe the motions of the heavenly bodies with accuracy matching reality sufficiently were developed (Figure 1.1). This is basically the

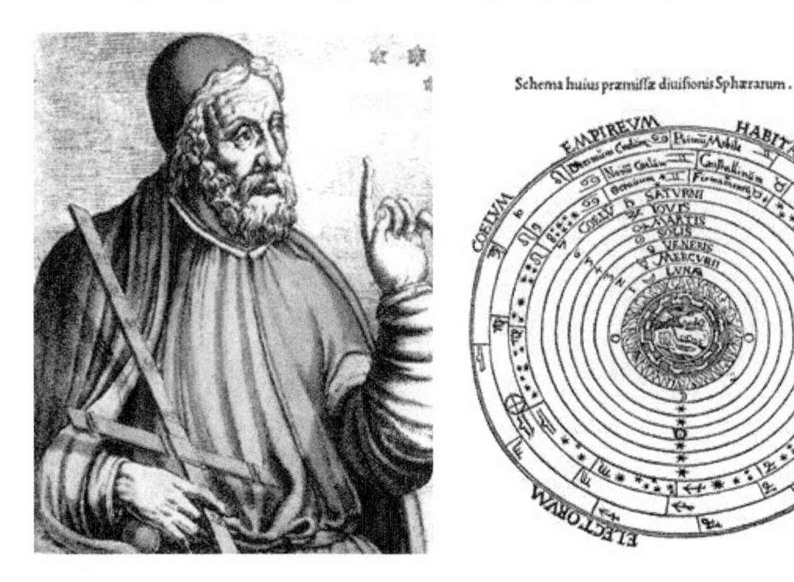


**FIGURE 1.1**
(Left) An image of Ptolemy; (Right) Ptolemy's model of our solar system

essence of numerical analysis. Numerical analysis is concerned with obtaining approximate solutions to problems while maintaining reasonable bounds of error, because it is often impossible to obtain exact answers. Numerical analysis makes use of algorithms to approximate solutions. Model development and solving the models is important to the world, for example in astronomy, construction, agriculture, architecture, and, of course, in engineering! In chemical engineering we use models and their (numerical) solutions to describe reactors and separators (dynamic and steady state), to perform computational fluid dynamics, to solve thermodynamic equations of state, to optimize process performance, to design and synthesize processes, and to regress experimental data, e.g., isotherms, kinetics, and so forth.

## 1.2 The idea of a model

In Figure 1.2 we can see an image by the Belgian surrealist Rene Magritte. It is a pipe, and below this pipe is a sentence in French that says, “Ceci n’est pas une pipe” (“this is not a pipe”). Actually, it is, indeed, not a pipe; it is an

**FIGURE 1.2**
“The Treachery of Images” by Rene Magritte

| Type of model | Criterion or classification |
|---|---|
| Mechanistic | Based on mechanisms/underlying phenomena (first principles) |
| Empirical | Based on input-output data, trials or experiments |
| Stochastic | Contains elements that are probabilistic in nature |
| Deterministic | Based on cause–effect analysis |
| Lumped parameter | Dependent variables not a function of spatial position |
| Distributed parameter | Dependent variables as a function of spatial position |
| Linear | Superposition applies |
| Nonlinear | Superposition does not apply |
| Continuous | Dependent variables defined over continuous space |
| Discrete | Only define for discrete values of time and/or space |
| Hybrid | Containing continuous and discrete behavior |

**TABLE 1.1**
Model types and their classifications

image of a pipe. Models are similar. Models are not the reality, they are an approximate description of reality. Eykhoff [20] defines an engineering model as a representation of the essential aspects of an existing system (or a system to be constructed) which presents knowledge of that system in a usable form. This implies basically that a model is (always) a simplification of reality. A model as such can give insight into the behavior of the system under study, but it does not always mean that this insight is phenomenological. For example, if an engineer develops a controller for a distillation tower, he would like to know how the distillation tower behaves dynamically. Whether this knowledge is based on first principles or not is not really relevant for his purposes. In Table 1.1 the different model types are listed.

The mathematical forms of the different model types can involve linear algebraic equations, nonlinear algebraic equations, ordinary differential equations, differential algebraic equations and partial differential equations. Each of the equation forms requires special techniques for solution.

## 1.3 Model building

Although there have been many attempts to structure the process of setting up process models to describe phenomena or systems, the general notion is that each modeling problem requires a custom-made approach. The applications and requirements are so different that general model development strategies would be extremely difficult and decisions regarding the modeling of a system can often best be made by an expert. However there is some kind of agreement on the four elementary steps in the modeling process: problem definition, design, evaluation and application.

In the problem definition phase, the modeling problem and the goal of the model are properly formulated. This formulation is based on performance and structure requirements with respect to the application and on the modeling expertise of the modeler. In the design phase, the structure and key variables of the model are identified.

For mechanistic models, the structure of the model reflects the physical structure of the system. This often means that additional steps have to be taken such as the formulation of physical and chemical laws and a proper translation of the major assumptions made in the design stage.

Key in the process is the application of conservation principles for conserved extensive quantities. Another important component is the development of constitutive relations, which are normally used to complete the model.

The conservation principle holds for mass, energy and momentum, and states that these quantities are neither destroyed or created but simply change form. Conservation principles lead to typical gas-liquid-solid systems involving the mass, component and energy balances. In particulate systems, particle number balances for the generation of population balance equations are also considered.

Constitutive relations are normally algebraic equations. Constitutive equations describe five classes of relations in a model, the mass, energy/heat transfer, the reaction rate expressions, the so-called property relations (thermodynamic constants and relations), the balance volume relations which define the connections between mass and energy, and the equipment and control constraints. For a detailed description of conservation and constitutive equations the reader is referred to Hangos and Cameron [25].

In the evaluation phase the model is verified with respect to its structure and the results of the model are validated with the real world situation. In this phase also the requirements with respect to model structure as formulated in the problem definition phase are evaluated. If all criteria are satisfied, the model can be applied.

## 1.4 Model analysis

The analysis or evaluation of models concerns two parts: numerical performance validation and model structure verification. Validation and verification are two strictly different criteria for model analysis.

Verification is related to the mathematical correctness of the model structure. Using commonsense is of importance, but there are also instruments available which aid the verification process. For example, one could study the range in which the parameters of the model are valid, or one could study the sensitivity of the parameters in relation to the outputs of the model. Models should be kept as simple as possible. If one model contains more parameters than another its performance can be better but it is also more complex. There are structure tests available that quantify the model structure on the basis of modeling error and model complexity. One of these tests is, for example, Akaike's information criterion.

In model validation the outputs of the model are compared with actual measurements from the system to determine whether the model describes the real system adequately. Cross-parameter validation, residual tests, and the root mean squared error are a few examples of model validation instruments. Chapter 11 on data regression will discuss model validation in more detail.

## 1.5 Model solution strategies

The main objective of this book is to prime the reader with the actual numerical solution strategies to the formulated process models. In Table 1.2 common solution approaches are listed for different equation systems.

| Equation system | Solution strategy |
|---|---|
| Linear | Matrix inversion |
| Ch. 2,3,4 | Gaussian elimination |
| | Jacobi method |
| Nonlinear | Newton's method |
| Ch. 5 | Secant method |
| | Broyden's method |
| ODEs | Euler's method (implicit/explicit) |
| Ch. 6 | Runge Kutta method |
| PDEs | Method of lines |
| Ch. 7,8 | Finite Volume Method |

**TABLE 1.2**
Model types and solution strategies

## 1.6 The seven-step modeling procedure

In setting up models for engineering application, a good practice is to follow a well-described sequence of modeling steps. Several modeling methodologies have been proposed, but here we will follow the seven-step procedure discussed in Cameron and Hangos.

You should understand that development of the model through a sequence of steps often leads to iterations—moving up and down through the modeling steps—especially when unusual or unwanted developments occur later in the process. No one gets it right the first time.

Before starting the setup of a process model, the problem definition should be clearly stated. This defines the process, the modeling goal, and the validation criteria. These decisions precede the seven-step modeling procedure, which is given in the form of an algorithmic problem. An algorithm is a systematic procedure for carrying out the modeling task. Figure 1.3 shows the systematic modeling procedure of Cameron and Hangos.

For an algorithmic problem, you should formally specify the following items:

1. The inputs to the problem should be listed in the "Given" section.
2. The desired output of the procedure should be listed in the "Find" or "Compute" section.
3. The method description should be listed in the "Procedure" or "Solution" section.

By following the above principles, the following algorithmic problem statement can be constructed.

The seven-step modeling procedure:

Given:

1. A process system
2. A modeling goal
3. Validation criteria
4. A mathematical model

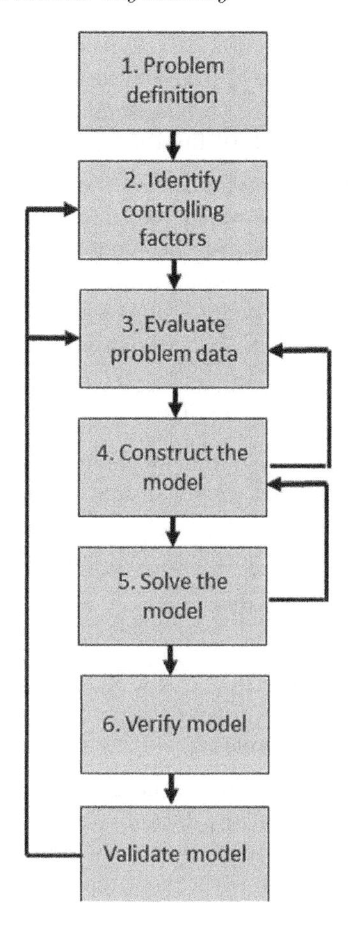


**FIGURE 1.3**
Systematic model building steps

The seven steps for model development:

*Step 1: Define the problem*

In Step 1, all relevant inputs and outputs have to be fixed. The type of spatial distribution has to be defined (i.e., distributed or lumped models), and the necessary range and accuracy of the model need to be specified. Additionally, the time characteristics of the model should be identified (i.e., stationary or dynamic).

*Step 2: Identify the controlling factors or mechanisms*

The next step is to explore the physicochemical processes and phenomena taking place in the system that are related to our modeling goal. These are called the controlling factors or mechanisms. The most common controlling factors include (1) chemical reaction, (2) diffusion, (3) conduction of heat, (4) forced convection heat transfer, (5) free convection heat transfer, (6) radiation heat transfer, (7) evaporation, (8) turbulent mixing, (9) heat or mass transfer through a boundary layer, and (10) fluid flow.

When modeling, we must understand that there is a set of process characteristics that are never fully identified. We often only identify and include a subset of essential characteristics in the model, which means that some essential characteristics might be missing from our model description. It could also very well happen that we include unnecessary complexity in our model or include process characteristics that are actually not part of the system.

The issues sketched above are not easy to deal with and depend strongly on our understanding of the system. Model validation is key here.

One cannot take all possible controlling mechanisms into account, and for that reason, a filter should be applied when considering the following key elements:

1. The hierarchy level(s) relevant to the model
2. The type of spatial distribution
3. The necessary range and accuracy
4. The time characteristics

*Step 3: Evaluate data for the problem*

Models of real process systems are of the gray-box type: we almost always need to use either measured process data directly or estimate parameter values in our models. It is important to consider both measured process data and parameter values together with their uncertainties or precision. Typically, industrial measured data has uncertainties between 10 and 30#, while laboratory or pilot plant data has uncertainty ranges of 5%–20%, but, for example, reaction kinetic data can have uncertainties up to 500%!

In Step 3, we need to look at the data required for modeling, found in literature or measured directly. The outcomes of this data evaluation might lead us back to the decisions in Steps 1 and 2.

*Step 4: Develop a set of model equations*

The equations in a process model are either differential (partial as well as ordinary) or algebraic in nature. Differential equations originate from conservation balances; therefore, they can be termed balance equations. Algebraic equations are usually of mixed origin and are generally called constitutive equations.

*Step 5: Find and implement a solution procedure*

After setup of the mathematical model, we must identify its mathematical form and find or implement a solution procedure. In all cases we must make sure that the model is well posed; that is, excess variables or all degrees of freedom are satisfied. We must also try to avoid numerical problems, such as high index systems. Lack of solution techniques may prevent a modeler using a particular type of process model, and that can lead to additional simplifications. This could be the case with distributed parameter process models.

*Step 6: Verify the model solution*

Verification is determining whether the model is behaving correctly. Is it coded correctly and giving you the intended answer? This is not the same as model validation, where we check the model against reality. You need to check carefully that the model is correctly implemented. One aspect is the correct programming practice, where a top-down algorithm design can help out. The other component is to verify whether the model exhibits the right qualitative characteristics; for example, if modeling a reactor, an increase in reactor temperature should deliver an increase in conversion.

*Step 7: Validate the model*

Model validation concerns the determination of the quality of the model as compared to independent observations or assumptions. Often only a partial validation is carried out in practical cases, depending on the modeling goal.

There are several possibilities to validate a process model. The actual validation method strongly depends on the process system, the modeling goal, and the possibilities of getting independent information for validation. These possibilities include:

1. Experimentally verify the assumptions.
2. Compare the model behavior with the process behavior.
3. Develop analytical models for simplified cases and compare behavior.
4. Compare with other models using a common problem.
5. Compare the model directly with process data.

The tools to carry out this task include the use of sensitivity analysis to identify the key controlling inputs or system parameters as well as the use of statistical validation tests. They can involve hypothesis testing and the use of various measures such as averages, variances, maxima, minima, and correlation factors.

If the validation results show that the developed model is not suitable for a modeling goal, then one has to return to Step 2 and perform the sequence again. Usually validation results indicate how to improve the model. We can often identify inadequate areas in our model development, and for that reason, not all modeling efforts are lost. Once more, it is mentioned that often one has to move up and down through the seven-step modeling hierarchy in an iterative fashion.

## 1.7 Ingredients of process models

The modeling procedure described in the previous section demonstrates that a model resulting from this procedure is not simply a set of equations. It incorporates a lot more information. To encourage clarity of presentation and consistency of the process model, a structural presentation incorporating all key ingredients is suggested.

List the:

1. Assumptions: time characteristics, spatial characteristics, flow conditions, controlling mechanisms or factors, neglected dependencies (such as temperature or concentration dependency of physicochemical properties), and the required range of states and associated accuracy.
2. Model equations and characterizing variables: the differential (balance) equations for
   a. overall mass, component masses for all components for all phases in all equipment at hierarchical levels, energy (or enthalpy) and momentum
   b. constitutive equations for mass transfer, heat/energy transfer and reaction rates, property relations (thermodynamic relations for equilibrium, temperature pressure, and composition), balance volume relations (relationships between mass and energy balance regions), and equipment and control equations.
3. Initial conditions: needed for differential balance equations in dynamic process models.
4. Boundary conditions: must be specified for differential balance equations in spatially distributed process models.
5. Parameters: the values and/or sources of the parameters have to be specified with their units and precision.

## 1.8 Summary

In this chapter the role of models in chemical engineering was briefly discussed. First, the basic definition of a model was introduced. Subsequently a common way of model building was discussed, starting with a problem definition phase, followed by design, evaluation, and application. After the model building phase, model analysis is done in two ways: the mathematical correctness of the model (model verification) and the numerical performance of the model (model validation). In the final section, a short overview of model solution strategies was given.

## 1.9 Exercises

### Exercise 1

Describe the major steps to be taken when building a model of a process system, and explain the reason for each step.

### Exercise 2

Explain the differences between stochastic, empirical, and mechanistic models. Describe which factors make it difficult or easy to develop these models?

### Exercise 3

Review what kinds of models are used in a particular industry sector (select, for example, the food, petrochemical, or pharmaceutical industries) and why they are used. Discuss how the modeling efforts relate to the potential benefits derived from their use?

# 2

# *Errors in computer simulations*

## 2.1 Introduction

Just like real experiments, computer simulations may contain errors. The big question is: How good are the results of computer simulations? It is imperative to understand the idea that errors occur in computer simulations and that the outcomes of simulation studies should always be checked.

To create awareness of how errors influence the outcomes of computer simulations, this chapter is used to introduce the types of errors that generally may be encountered and how these errors may influence simulation results.

In principle, errors in simulations can be categorized into five classes: 1) errors in the mathematical model, 2) errors in the input data, 3) errors in computer programs, 4) round-off and truncation errors, and 5) break errors.

Errors in the formulation of the mathematical model are not discussed in this chapter, but it should be realized that, in principle, each model is a simplification of reality; leaving physical or chemical effects out of the model will increase robustness, but will probably decrease accuracy. The other types of errors will be discussed in the following sections.

## 2.2 Significant digits

It may be clear that the result $x$ is not calculated by the computer, but is just an approximation $\tilde{x}$. This approximation only has value if we have an idea of how big the difference is, or the *absolute error*, $\delta$, given as:

$$\delta = \tilde{x} - x, x \neq 0. \tag{2.1}$$

For all $x$, except $x = 0$, we may also speak of the *relative error*:

$$\frac{\tilde{x} - x}{\tilde{x}}. \tag{2.2}$$

Of course, the exact value of $\delta$ is unknown, and, generally, estimations of the error are used, which are formulated as a relative or absolute *error margin.* If $\delta$ is an absolute error, we express the error margins as:

$$\tilde{x} - \delta \leq x \leq \tilde{x} + \delta, \tag{2.3}$$

or simply as

$$x = \tilde{x} \pm \delta. \tag{2.4}$$

Instead of relative error, we often use the number of *significant digits.* $\tilde{x}$ has $m$ significant digits when the absolute value of the error in $\tilde{x}$ is smaller or equal to 5 on the $(m+1)^{th}$ position:

$$10^{q-1} \leq |\tilde{x}| \leq 10^{q} \tag{2.5}$$

$$|x - \tilde{x}| \leq 0.5 \times 10^{q-m}. \tag{2.6}$$

For example, consider $x = \frac{1}{3}$ and the approximation $\tilde{x} = 0.333$. The absolute error margin in $\tilde{x}$ equals $\frac{1}{3} - 0.333 = 0.000333\cdots$ This is smaller than 0.0005, so $m + 1 = 4$, meaning $m = 3$. In other words, $\tilde{x}$ has three significant digits.

## 2.3 Round-off and truncation errors

If you perform calculations with a computer or a calculator, you will have to deal with the representation of numbers. Computers can represent numbers with a large, but finite number of digits. Each number is approximated by round-off and/or truncations.

As we are mostly using the decimal system, this means that for a digit $c$ that has position $n$, the value is represented to $c10^{n-1}$.

For example, the number 4521. The number 5 is in the third position and adds $5 \times 10^{3-1}$ to the total.

It is very well possible to use a different basis than the decimal one, for example — very commonly used in computers — the binary basis; 0 and 1: electrical current flows or it does not flow. Current corresponds to digits!

If you write the number 4521 as a binary number, you would end up with $1*2^{12}+0*2^{11}+0*2^{10}+0*2^{9}+1*2^{8}+1*2^{7}+0*2^{6}+1*2^{5}+0*2^{4}+1*2^{3}+0*2^{2}+0*2^{1}+1*2^{0}$, or shortly 1000110101001. To prevent misunderstandings, you could explicitly mention the basis, like $(4521)_{10} = (1000110101001)_2$, or in a more general way:

$$(c_m \cdots c_1 c_0)_q = c_0 q^0 + c_1 q^1 + \cdots + c_m q^m, c \in 0, 1, 2, \cdots, q-1. \tag{2.7}$$

Numbers are stored as segments in the computer memory, usually called *words.*

We distinguish two types of numbers, namely the *integers* and the *floating point numbers.*

Natural numbers are represented by computers as integers, so the integer representation for binary numbers can be given as:

$$z = \sigma(c_0 2^0 + c_1 2^1 + \cdots + c_{\lambda-1} 2^{\lambda-1}), \tag{2.8}$$

where $\sigma$ is the sign of $x$; for example, when negative, $\sigma = 0$ and when positive $\sigma = 1$. The digit $c_i$ can take the values of 0 and 1 for every $i = 1, 2, \cdots, \lambda - 1$, where $\lambda$ is the length of the word.

As an example, in Fortran, integers are represented by 32-bit words, so $\lambda = 31$. The whole set of integers available for $z$ is:

$$-2^{31} + 1 \leq z \leq 2^{31} \approx 2 * 10^9. \tag{2.9}$$

If during calculations the number becomes bigger than $2^\lambda - 1$, the computer reports an *overflow*; of course, the same message appears if it becomes smaller than $-2^\lambda + 1$.

Real numbers are represented by the computer as floating point numbers. On a binary basis a real number $x$ can be represented as

$$x = \sigma(2^{-1} + c_2 2^{-2} + \cdots + c_m 2^{-m}) 2^e, \tag{2.10}$$

where $\sigma$ is the sign of $x$ (positive or negative) and $c_i (i = 1, 2, 3, \cdots, m)$ can have the value 0 or 1. The exponent $e$ is always an integer.

The part between brackets is called the *mantissa*. Notice that $c_1$ is always 1. We have to choose the exponent $e$ in such a way that the first nonzero digit corresponds to $2^{e-1}$.

The total number of available positions, $\lambda + 1$, has to be divided: the sign requires one position, the mantissa requires $m - 1$ positions, so the exponent has only $\lambda - m + 1$ positions. Ergo, the maximum value for the exponent equals $2^{\lambda-m} - 1$; the minimum value equals $-2^{\lambda-m} + 1$.

Just as with the integer representation, there is an upper and lower margin, and values outside these margins cannot be represented.

Let us have a look at an example: take $\lambda = 3$, $m = 2$, and $x = \frac{2}{3}$. All representable numbers are of the form $\pm(2^{-1} + c_2 2^{-2}) 2^e$, where $c_2$ can only be 0 or 1 and $e = \pm a_0 2^0$. $a_0$ can only be 0 or 1.

There are two options to represent $x = \frac{2}{3}$, namely by *truncation*: $2^{-1}$, or by *round-off*: $2^{-1} + 2^{-2}$. Such inaccuracies are often not acceptable, and for this reason most computers offer the possibility to represent numbers with double precision, where the length of the mantissa is doubled.

## 2.4 Break errors

Computer calculations cannot, of course, take infinite time. At some point, calculations need to be interrupted. This interruption inevitably leads to error, often called *break error.*

To illustrate the phenomenon of break error, we can consider the calculation of $e^x$ by a Taylor series:

$$e^x = \sum_{n=0}^{\infty} \frac{x^n}{n!} = \frac{x^0}{0!} + \frac{x^1}{1!} + \frac{x^2}{2!} + \cdots \tag{2.11}$$

By using a computer, we can summate a large, but finite number of Taylor series terms, for example $N + 1$ terms:

$$e^x \approx \sum_{n=0}^{N} \frac{x^n}{n!} = \frac{x^0}{0!} + \frac{x^1}{1!} + \frac{x^2}{2!} + \cdots + \frac{x^N}{N!}. \tag{2.12}$$

The difference between the left- and right-hand terms of the $\approx$ sign is called the *break error.*

## 2.5 Loss of digits

In principle, all numerical algorithms can be composed of four basic operations: adding, subtracting, multiplication, and division. However, computers cannot perform these operations without any error.

For example, the division $x/y$ will not be exact, even not when the operands $x$ and $y$ are without error.

If we assume that $\tilde{x}$ and $\tilde{y}$ contain errors $\delta$ and $\epsilon$, we may write the absolute error margins as

$$\tilde{x} - \delta \le x \le \tilde{x} + \delta, \tilde{y} - \epsilon \le y \le \tilde{y} + \epsilon. \tag{2.13}$$

The sum $x + y$ equals, of course,

$$(\tilde{x} + \tilde{y}) - (\delta + \epsilon) \le x + y \le (\tilde{x} + \tilde{y}) + (\delta + \epsilon), \tag{2.14}$$

So the absolute error of $x + y$ equals $\delta + \epsilon$. Also, the absolute error for subtraction equals $\delta + \epsilon$. The same counts for multiplication and division.

Let us evaluate an example, to see what will happen to the error if two numbers are subtracted that have more or less the same values: $x = \pi$ and $\tilde{x} = 3.1416$, $y = \frac{22}{7}$ and $\tilde{y} = 3.1429$. The absolute error margins are $\delta = 7.35 * 10^{-6}$ and $\epsilon = 4.29 * 10^{-5}$. The relative errors of $\tilde{x}$ and $\tilde{y}$ are $2.34 * 10^{-6}$ and $1.36 * 10^{-5}$, respectively. But the relative error of $\tilde{x} - \tilde{y}$ is much larger: 0.028. The relative error increases strongly! $\tilde{x}$ and $\tilde{y}$ both have 5 significant digits, but the difference $\tilde{x} - \tilde{y}$ only has 2 significant digits: by performing the subtraction, significant digits are lost!

Another example can be encountered if $e^{-5}$ is calculated, using Equation 2.14. If you calculate $e^{-5}$ without error, you will find a value of 0.006738. But, if you only use numbers fixed on 4 digits for your calculation, you can observe the results in the following table:

| n | $1 + (-5)^0/1! + ... + (-5)^n/n!$ |
|---|---|
| 0 | 1.000 |
| 1 | -4.000 |
| 2 | 8.500 |
| 3 | -12.33 |
| 4 | 13.71 |
| 5 | -12.33 |
| 6 | 9.38 |
| ... | ... |
| 23 | 0.009989 |
| 24 | 0.009989 |
| 25 | 0.009989 |

For higher values of $n$ you find 0.009989, something remarkably different as the real value. The continuously changing positive and negative sign is the cause of the loss of digits in this problem.

You could prevent loss of digits, by first calculating $e^1$ and subsequently calculating $1/(e * e * e * e * e)$. The Taylor series of $e^1$ does not alternate, so loss of digits is prevented!

## 2.6 Ill-conditioned problems

The introduction already mentioned is that errors in simulations can be traced back to 5 sources. One of these sources is the error in output as a result of errors in the input information; see Figure 2.1. This type of error is called the *propagated error*. If the input $x$ has a small error $\delta x$, then the output will be $f(x + \delta x)$ instead of $f(x)$. The propagated error is in this case $f(x + \delta x) - f(x)$. Often we cannot obtain a good estimation of $f(x + \delta x)$. For this reason, the best way to describe the propagated error is by means of a Taylor series

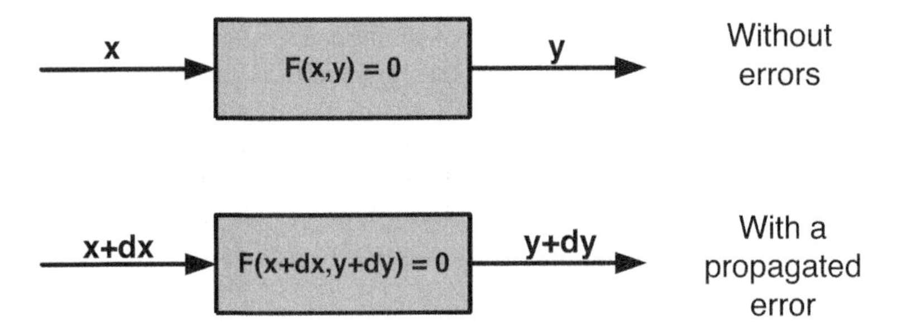


**FIGURE 2.1**
Input–output representation without and with errors

expansion of $f$ around $x$:

$$f(x + \delta x) - f(x) \approx \delta x f'(x). \tag{2.15}$$

Another way of quantifying a propagated error is by means of the so-called condition criterion $C$, defined as the ratio of relative error in the output and relative error in the input:

$$C = \max_{\delta x} \left( \frac{\delta y / y}{\delta x / x} \right). \tag{2.16}$$

If $C \leq 10$, the propagation of an error will be small.

We can illustrate error propagation with an example. We would like to solve the linear system $Ay = x$, with

$$A = \begin{bmatrix} 1.2969 & 0.8648 \\ 0.2161 & 0.1441 \end{bmatrix},$$

and $x = [0.8642, 0.1440]$.

The solution of this problem is $y = [2, -2]$. However, by fixing the input data $x$ on 11 decimals and subsequently performing Gaussian elimination, we obtain as a solution $y = [0.662, -0.0002]$.

Apparently this system is *ill conditioned*; a small error in the inputs result in a big error in the outputs. If you calculate the condition criterion, you would find $C \approx 2 * 10^8$.

Also, for bigger systems, small errors in the inputs can produce tremendous errors in the solution vectors. Such badly conditioned systems are not easy to solve accurately. Sometimes it is possible to reformulate the problem in order to improve the conditioning. Sometimes the solution method can be improved.

## 2.7 (Un-)stable methods

The condition criterion tells something about the sensitivity of the solution, but is does not tell you anything about the quality of the solution method you are using. If a solution method propagates the error, we call the method *unstable.*

This is best illustrated with an example. Consider the following recurrent relationship:

$$y_{n+1} = y_{n-1} - y_n \tag{2.17}$$

with $y_0 = 1$, $y_1 = 2/(1+\sqrt{5})$. You can prove by substitution that:

$$y_n = x^{-n}, \tag{2.18}$$

with $n = 1, 2, \cdots$ and $x = (1+\sqrt{5})/2$. If we use Equations 2.17 and 2.18 to calculate numbers starting with $n = 1$ and ending with $n = 30$, we will obtain the following numbers

| $n$ | $y_n$ | $x^{-n}$ |
|---|---|---|
| 1 | 1.000 | 1.000 |
| 2 | 0.6180 | 0.6180 |
| 3 | 0.3819 | 0.3819 |
| 4 | 0.2361 | 0.2361 |
| ... | ... | ... |
| 25 | $0.1237 * 10^{-2}$ | $0.5960 * 10^{-5}$ |
| 26 | $-0.1989 * 10^{-2}$ | $0.3684 * 10^{-5}$ |
| 27 | $0.3226 * 10^{-2}$ | $0.2276 * 10^{-5}$ |
| 28 | $-0.5215 * 10^{-2}$ | $0.1407 * 10^{-5}$ |
| 29 | $0.8442 * 10^{-2}$ | $0.8689 * 10^{-6}$ |
| 30 | $-0.1365 * 10^{-1}$ | $0.5374 * 10^{-6}$ |

The differences are striking. The results obtained with Equation 2.17 are not correct for larger values of $n$. And it seems that the results also alternate from positive to negative. This oscillating behavior is a strong indicator for the fact that the calculation method of Equation 2.17 is unstable.

If you take a closer look at the error propagation you can deduce the following relationship:

$$\tilde{y}_n = y_n - \left[\left(\frac{1+\sqrt{5}}{2}\right)^{-n} - \left(\frac{1-\sqrt{5}}{2}\right)^{-n}\right]\frac{6\delta}{\sqrt{5}}, \tag{2.19}$$

where $\delta$ is the computation error. The second part of the equation especially, will lead to error; as $n$ increases, the second term will also grow, no matter how small your $\delta$ is.

In conclusion, the method (or problem formulation) that is chosen for calculation may strongly influence the end result.

## 2.8 Summary

In this chapter, we found that the number of significant digits determines the accuracy of a number. Computers are limited in the expression of numbers, and can principally represent numbers in two ways: as integers and as floating point numbers. On the basis of word length, computers round off or truncate numbers, which may lead to error. Numerical operations may lead to loss of significant digits, resulting in computational error. Also, errors in input data may propagate the error in the output data, and repeated operations may increase the error. The formulation of the problem and the method that is used to solve the problem determine if the problem is conditioned properly and/or if the method is stable or not.

## 2.9 Exercises

### Exercise 1

The numbers $x_1 = 514.01, x_2 = -0.04518, x_3 = 23.4604$ are approximated by $\hat{x}_1 = 514.023, \hat{x}_2 = -0.045113, \hat{x}_3 = 23.4213$. Calculate the number of significant digits in $\hat{x}_1, \hat{x}_2$ and $\hat{x}_3$.

### Exercise 2

Determine the binary representation of the following for numbers, given the decimal system: 129, 0.1, 0.2 and 0.8125.

Change the following binary numbers into their decimal equivalents: $(1111111111)_2$, $(10101.101)_2$ and $(.101010101...)_2$.

### Exercise 3

We would like to calculate the value of the function

$$f(x) = 1 - x^2 \tag{2.20}$$

for a given value of $x \in (0, 1)$. Calculate the condition criterion $C$ of this problem.

For which values of $x \in (0, 1)$ does it hold that $C > 10$? What does that mean for the propagated error?

# 3

# *Linear equations*

## 3.1 Introduction

In Chapter 3 we are going to write our first MATLAB program to solve systems of linear equations. The aim of this chapter is to familiarize you with some basic theory from linear algebra. We will see that systems of linear equations can be looked at in different ways. We will also list some basic ideas from matrix theory—the inverse, the determinant, and the rank. We will also look at eigenvalues and eigenvectors. But first something about MATLAB as a programming language.

## 3.2 MATLAB

MATLAB is a high-level computer programming language, so you don't need to worry about memory management/allocation issues. However, it is still close enough to "proper" programming languages. It has a command line interface and is quite user friendly. It is also important that vector and matrix computations are an intrinsic part of MATLAB. Because MATLAB is an interpreted language (in comparison to a compiler-based language like C++) it is slow for certain operations.

## 3.3 Linear systems

Linear equations can be written in several forms. For each case the solution to the equations has a different interpretation. We could, for example, write a linear system as *separate equations*:

$$x + y + z = 4 \quad (3.1)$$
$$2x + y + 3z = 7 \quad (3.2)$$
$$3x + y + 6z = 5, \quad (3.3)$$

or we could represent the system as a *matrix mapping*:

$$\begin{bmatrix} 1 & 1 & 1 \\ 2 & 1 & 3 \\ 3 & 1 & 6 \end{bmatrix} \begin{bmatrix} x \\ y \\ z \end{bmatrix} = \begin{bmatrix} 4 \\ 7 \\ 5 \end{bmatrix}, \quad (3.4)$$

or briefly as:

$$Mx = b, \quad (3.5)$$

where $M$ is the matrix and $x$ and $b$ are vectors. Sometimes a linear system is represented as *linear combinations of basis vectors*:

$$x \begin{bmatrix} 1 \\ 2 \\ 3 \end{bmatrix} + y \begin{bmatrix} 1 \\ 1 \\ 1 \end{bmatrix} + z \begin{bmatrix} 1 \\ 3 \\ 7 \end{bmatrix} = \begin{bmatrix} 4 \\ 7 \\ 5 \end{bmatrix}. \quad (3.6)$$

## 3.4 The inverse of a matrix

If we want to solve a linear system $Mx = b$, we need, in fact, the *inverse* of the matrix $M$, provided that the matrix is square. The inverse of a matrix is defined as

$$MM^{-1} = I \leftrightarrow M^{-1}M = I, \quad (3.7)$$

where $I$ is the identity matrix. If we multiply both sides of Equation 3.5 with $M^{-1}$ we get:

$$M^{-1}Mx = M^{-1}b. \quad (3.8)$$

If we now merge Equation 3.7 into 3.8 we obtain

$$Ix = M^{-1}b = x \quad (3.9)$$

or

$$x = M^{-1}b. \quad (3.10)$$

The question now is, of course, how to determine the inverse. The inverse can be found by

$$M^{-1} = \frac{1}{\det(M)} \begin{bmatrix} C_{11} & C_{12} & C_{13} \\ C_{21} & C_{22} & C_{33} \\ C_{31} & C_{32} & C_{33} \end{bmatrix}^T, \quad (3.11)$$

where $C_{ij}$ are the *co-factors* of the matrix $M$. Calculation of co-factors is best illustrated by an example. Consider our matrix $M$

$$M = \begin{bmatrix} 1 & 1 & 1 \\ 2 & 1 & 3 \\ 3 & 1 & 6 \end{bmatrix}. \tag{3.12}$$

## 3.5 The determinant of a matrix

If we want to calculate the co-factor of element $M_{11}$, we have to first calculate the *determinant* of the stuff that is left over when you cover up the row and column of the element in question, and thus, for element $M_{11}$ we consider

$$\begin{bmatrix} 1 & \star & \star \\ \star & 1 & 3 \\ \star & 1 & 6 \end{bmatrix} \tag{3.13}$$

and calculate

$$C_{11} = +\det \begin{bmatrix} 1 & 3 \\ 1 & 6 \end{bmatrix} = 6 \times 1 - 3 \times 1 = 3. \tag{3.14}$$

The plus sign comes from the following matrix:

$$\begin{bmatrix} + & - & + \\ - & + & - \\ + & - & + \end{bmatrix}.$$

After calculating the co-factors for each element in the matrix, the result is:

$$\begin{bmatrix} 1 & 1 & 1 \\ 2 & 1 & 3 \\ 3 & 1 & 6 \end{bmatrix}^{-1} = \frac{1}{\det(M)} \begin{bmatrix} 3 & -5 & 2 \\ -3 & 3 & -1 \\ -1 & -2 & -1 \end{bmatrix}. \tag{3.15}$$

If the inverse of a matrix does not exist, there are either no solutions or infinitely many solutions. The determinant determines the existence of an inverse. If the determinant is zero, an inverse does not exist, and the matrix is called *singular*. We can calculate $\det(M)$ by multiplying each element on a row by its co-factor and adding the result:

$$\det\left(\begin{bmatrix} 1 & 1 & 1 \\ 2 & 1 & 3 \\ 3 & 1 & 6 \end{bmatrix}\right) = +\det \begin{bmatrix} 1 & 3 \\ 1 & 6 \end{bmatrix} - \det \begin{bmatrix} 2 & 3 \\ 3 & 6 \end{bmatrix} + \det \begin{bmatrix} 2 & 1 \\ 3 & 1 \end{bmatrix} = -1. \tag{3.16}$$

Or you can do the same thing for columns:

$$\det\left(\begin{bmatrix} 1 & 1 & 1 \\ 2 & 1 & 3 \\ 3 & 1 & 6 \end{bmatrix}\right) = +\det\begin{bmatrix} 2 & 1 \\ 3 & 1 \end{bmatrix} - \det\begin{bmatrix} 1 & 1 \\ 3 & 1 \end{bmatrix} + \det\begin{bmatrix} 1 & 1 \\ 2 & 1 \end{bmatrix} = -1. \tag{3.17}$$

Now, we have everything to solve our problem:

$$\begin{bmatrix} x \\ y \\ z \end{bmatrix} = \frac{1}{-1}\begin{bmatrix} 3 & -5 & 2 \\ -3 & 3 & -1 \\ -1 & -2 & -1 \end{bmatrix}\begin{bmatrix} 4 \\ 7 \\ 5 \end{bmatrix} = \frac{1}{-1}\begin{bmatrix} -13 \\ 4 5 \end{bmatrix} = \begin{bmatrix} 13 \\ -4 \\ -5 \end{bmatrix}. \tag{3.18}$$

For large matrices, computation of determinants and inverses in this way is too difficult (slow), so we need other methods to calculate the inverse of a large matrix.

## 3.6 Useful properties

A triangular matrix holds that:

$$\det(M) = \prod_{i=1}^{n} a_{ii}. \tag{3.19}$$

If we want to calculate the determinant of:

$$M = \begin{bmatrix} 5 & 3 & 2 \\ 0 & 9 & 1 \\ 0 & 0 & 1 \end{bmatrix}, \tag{3.20}$$

we can easily do that with Equation 3.19

$$\det(M) = 5 \times 9 \times 1 = 45. \tag{3.21}$$

Another useful property is that the determinant of a matrix multiplication is equal to the product of the determinants of the individual matrices:

$$\det(AM) = \det(A) * \det(M). \tag{3.22}$$

We may use this rule to quickly calculate the determinant of the following matrix:

$$A = \begin{bmatrix} a & 0 & 0 \\ 0 & 1 & 0 \\ 0 & 0 & 1 \end{bmatrix}. \tag{3.23}$$

We can write out this matrix as a scalar $a$ and an identity matrix $M$, and subsequently use the rule of Equation 3.22 to find:

$$\det(AM) = \det(A) * \det(M) = a * \det(M). \tag{3.24}$$

## 3.7 Matrix ranking

The *rank* of a matrix is defined as the number of linearly independent columns, i.e., columns that cannot be expressed as a linear combination of the other columns of the matrix. By reducing a matrix to its upper triangular form we can easily identify how many linearly independent basis vectors there are.

Consider, for example,

$$M = \begin{bmatrix} 5 & 3 & 2 \\ 0 & 9 & 1 \\ 0 & 0 & 1 \end{bmatrix}. \tag{3.25}$$

This matrix has 3 independent columns, $rank(M) = 3$. One other example is as follows:

$$M = \begin{bmatrix} 1 & 2 & 1 & 0 \\ 0 & 0 & 1 & 1 \\ 0 & 0 & 0 & 0 \end{bmatrix}. \tag{3.26}$$

This matrix has 2 independent columns and 2 dependent columns, as column 2 can be expressed as 2 times column 1, and column 4 equals column 3 minus column 1. In this case $rank(M) = 2$.

A solution to a system of linear equations may or may not exist, and it may or may not be unique. The existence of solutions can be determined by comparing the rank of a matrix $M$ to the rank of an *augmented* matrix $M_a$. If we have a linear system $Mx = b$, where

$$M = \begin{bmatrix} a_{11} & a_{21} & a_{31} \\ a_{12} & a_{22} & a_{32} \\ a_{13} & a_{23} & a_{33} \end{bmatrix}$$

and

$$b = \begin{bmatrix} b_1 \\ b_2 \\ b_3 \end{bmatrix},$$

then the augmented matrix is

$$M_a = \begin{bmatrix} a_{11} & a_{21} & a_{31} & b_1 \\ a_{12} & a_{22} & a_{32} & b_2 \\ a_{13} & a_{23} & a_{33} & b_3 \end{bmatrix}.$$

When $rank(M) = n$, where $n$ is the size of the matrix, there exists a unique solution. When $rank(M) < n$ and $rank(M) = rank(M_a)$, there is an infinite number of solutions and when $rank(M) < n$ and $rank(M) < rank(M_a)$ there is no solution.

For example, the linear system $Mx = b$ with

$$M = \begin{bmatrix} 1 & 1 & 2 \\ 0 & 3 & 1 \\ 0 & 0 & 2 \end{bmatrix}, b = \begin{bmatrix} 17 \\ 11 \\ 4 \end{bmatrix}$$

has $rank(M) = 3$ and $n = 3$, so, there is a unique solution to this problem. Consider the following system:

$$M = \begin{bmatrix} 1 & 1 & 2 \\ 0 & 3 & 1 \\ 0 & 0 & 0 \end{bmatrix}, b = \begin{bmatrix} 17 \\ 11 \\ 0 \end{bmatrix}.$$

Here, $rank(M) = rank(M_a) = 2$, which is smaller than $n$, so there is an infinite number of solutions.

## 3.8 Eigenvalues and eigenvectors

Matrices have characteristic directions, called *eigenvectors*. An eigenvector $e$ is defined by:

$$Me = \lambda e. \tag{3.27}$$

When an eigenvector is multiplied by the matrix $M$, the result is the eigenvector itself. The scale constant $\lambda$ is called the eigenvalue. Any multiple of an eigenvector is also an eigenvector.

We can derive from Equation 3.27 that $Me - I\lambda e = 0$, or $(M - I\lambda)e = 0$. In order to find values for $\lambda$, the determinant $\det(M - I\lambda)$ should be zero. We may write for a matrix $M$ with $n = 3$:

$$\det(M - I\lambda) = \det\left(\begin{bmatrix} 1-\lambda & 0 & 0 \\ 0 & 1-\lambda & 0 \\ 0 & 0 & 1-\lambda \end{bmatrix}\right) = 0, \tag{3.28}$$

or

$$\det(M - I\lambda) = (1-\lambda) + 1\det\begin{bmatrix} 1-\lambda & 0 \\ 0 & 1-\lambda \end{bmatrix} - 0\det\begin{bmatrix} 0 & 0 \\ 1 & 1-\lambda \end{bmatrix}$$
$$+ 1\det\begin{vmatrix} 0 & 1-\lambda \\ 1 & 0 \end{vmatrix} = 0. \tag{3.29}$$

Writing Equation 3.29 out will yield

$$(1-\lambda)\left[(1-\lambda)^2 - 0\right] + [0 - 1(1-\lambda)] = 0. \tag{3.30}$$

This is a third-order polynomial, so $\lambda$ has three roots, $\lambda_1 = 0$, $\lambda_2 = 1$, and $\lambda_3 = 2$.

If a matrix is square, you could decompose it into a diagonal matrix of eigenvalues, multiplied by matrices that have columns made up of the eigenvectors. For

$$Me_1 = \lambda_1 e_1 \tag{3.31}$$
$$Me_2 = \lambda_2 e_2 \tag{3.32}$$
$$Me_3 = \lambda_3 e_3 \tag{3.33}$$

this would result in

$$M \begin{bmatrix} \vdots & \vdots & \vdots \\ e_1 & e_2 & e_3 \\ \vdots & \vdots & \vdots \end{bmatrix} = \begin{bmatrix} \vdots & \vdots & \vdots \\ e_1 & e_2 & e_3 \\ \vdots & \vdots & \vdots \end{bmatrix} \begin{bmatrix} \lambda_1 & 0 & 0 \\ 0 & \lambda_2 & 0 \\ 0 & 0 & \lambda_3 \end{bmatrix}, \tag{3.34}$$

or more compactly written as

$$MU = U\Lambda \leftrightarrow M = U\Lambda U^{-1}. \tag{3.35}$$

We will use this spectral decomposition later on.

## 3.9 Spectral decomposition

**Theorem 1** *If A is a normal matrix, it is possible to find a complete orthonormal set of eigenvectors even if the matrix has eigenvalues of multiplicity greater than 1; i.e., $det(A - \lambda I) = 0$ has repeated roots. The matrix W whose columns are these eigenvectors is unitary, and we can write A as:*

$$A = W\Lambda W^H, \tag{3.36}$$

where $\Lambda = diag(\lambda_1, \lambda_2, \cdots, \lambda_N)$.

*Proof*: We first write A as a Schur decomposition:

$$A = URU^H. \tag{3.37}$$

Now, taking the Hermitian conjugate,

$$A^H = (URU^H)^H = UR^H U^H, \tag{3.38}$$

we then form the two matrix products:

$$AA^H = URU^H(UR^H U^H) = URR^H U^H \tag{3.39}$$
$$A^H A = UR^H U^H(URU^H) = UR^H RU^H. \tag{3.40}$$

For $A$ to be normal, $AA^H$ must be normal as well, $RR^H = R^H R$. For

$$R = \begin{bmatrix} R_{11} & R_{12} & R_{13} & \cdots & R_{1N} \\ & R_{22} & R_{23} & \cdots & R_{2N} \\ & & R_{33} & \cdots & R_{3N} \\ & & & \ddots & \vdots \\ & & & & R_{NN} \end{bmatrix} \quad (3.41)$$

and

$$R^H = \begin{bmatrix} \bar{R}_{11} & & & & \\ \bar{R}_{12} & \bar{R}_{22} & & & \\ \bar{R}_{13} & \bar{R}_{23} & \bar{R}_{33} & & \\ \vdots & \vdots & \vdots & \ddots & \\ \bar{R}_{1N} & \bar{R}_{2N} & \bar{R}_{3N} & \cdots & \bar{R}_{NN} \end{bmatrix}, \quad (3.42)$$

$RR^H = R^H R$ only if $R$ is diagonal. As $R$ is similar to $A$,

$$R = \Lambda = \begin{bmatrix} \lambda_1 & & & \\ & \lambda_2 & & \\ & & \ddots & \\ & & & \lambda_N \end{bmatrix}. \quad (3.43)$$

The Schur decomposition for a normal matrix is, therefore,

$$A = U \Lambda U^H. \quad (3.44)$$

Postmultiplication by $U$ yields

$$AU = U \Lambda. \quad (3.45)$$

The general form of the eigenvector decomposition is ($AW = W\Lambda$), where $W$ is a matrix whose column vectors are eigenvectors of A. Therefore, for any normal matrix $A$, we can form a unitary matrix whose column vectors are eigenvectors to write $A$ in Jordan normal form,

$$A = W \Lambda W^H. \quad (3.46)$$

For a matrix to be unitary, its column vectors must be orthogonal, as

$$\begin{aligned} W^H W &= \begin{bmatrix} - & \mathbf{w}^{(1)^H} & - \\ & \vdots & \\ - & \mathbf{w}^{(N)^H} & - \end{bmatrix} \begin{bmatrix} | & & | \\ \mathbf{w}^{(1)} & \cdots & \mathbf{w}^{(N)} \\ | & & | \end{bmatrix} \\ &= \begin{bmatrix} \mathbf{w}^{(1)} \cdot \mathbf{w}^{(1)} & \cdots & \mathbf{w}^{(1)} \cdot \mathbf{w}^{(N)} \\ \vdots & & \vdots \\ \mathbf{w}^{(N)} \cdot \mathbf{w}^{(1)} & \cdots & \mathbf{w}^{(N)} \cdot \mathbf{w}^{(N)} \end{bmatrix}. \end{aligned} \quad (3.47)$$

Therefore, it is always possible, for any normal matrix A, to find a complete, orthonormal basis for $C^N$ whose members are eigenvectors of $A$. One can write any vector $\nu \in C^N$ as the *spectral decomposition*

$$\nu = c_1\mathbf{w}^{(1)} + c_2\mathbf{w}^{(2)} + \cdots + c_N\mathbf{w}^{(N)} \quad (3.48)$$

$$A\mathbf{w}^j = \lambda_j\mathbf{w}^{(j)}\mathbf{w}^{(j)} \cdot \mathbf{w}^{(k)} = \delta_{jk}, \quad (3.49)$$

where $\mathbf{w}^{(j)} \in C^N$ and $c_j = \mathbf{w}^{(j)} \cdot \nu$.

## 3.10 Summary

In this chapter we found that linear equations can be written as matrices. Whether or not a solution exists, depends on the rank of a matrix. We also showed briefly what eigenvectors and eigenvalues are. Such matrix properties are useful in determining whether a system can be solved, or if a system is stable or not.

## 3.11 Exercises

### Exercise 1.a

The following linear system is given:

$$\begin{aligned} 2x_1 + x_2 + x_3 &= 4 \\ x_1 + 2x_2 + 2x_3 &= 3 \\ x_1 - x_2 + 6x_3 &= 1. \end{aligned}$$

Rewrite this system in terms of $Ax = b$ and then determine $A^{-1}$ with use of cofactors. Subsequently, solve the system with $A^{-1}b$.

### Exercise 1.b

Given is the system $Ax = b$, with

$$A = \begin{bmatrix} 1 & 1 & 0 \\ 2 & 1 & 1 \\ 1 & 0 & 1 \end{bmatrix}.$$

Prove that $A$ is singular. Find a $b$ for which this system does not have a solution, and find a $b$ for which $b$ has an infinite number of solutions.

### Exercise 2

Calculate the eigenvalues of the following matrices:

$$A = \begin{bmatrix} 1 & 1 & 0 \\ 0 & -1 & 0 \\ 0 & 0 & 2 \end{bmatrix}; B = \begin{bmatrix} 1 & 2 & 3 \\ 2 & 3 & 1 \\ 3 & 2 & 1 \end{bmatrix}.$$

### Exercise 3

Given this matrix:

$$A = \begin{bmatrix} -2 & 2 & -1 \\ 7 & 3 & -1 \\ -4 & -4 & -2 \end{bmatrix},$$

prove that the characteristic equation is given by: $-\lambda^3 - \lambda^2 + 30\lambda + 72 = 0$. Subsequently, determine $\lambda$ and give the eigenvectors of $A$ with the computer.

# 4

# *Elimination methods*

## 4.1 Introduction

In this chapter, we are going to write our first MATLAB program. This program can solve a set of linear equations. The method that we are going to use to perform the required row operations is called *Gaussian elimination*. But we will encounter some problems with Gaussian elimination, and for that reason we will resort to a decomposition technique called *LU factorization*.

## 4.2 MATLAB

We already introduced MATLAB in the previous chapter as a user-friendly programming language. When you open the MATLAB user interface you will see a division of three screens: a *work space*, a *command prompt*, and a *command history*. You can type commands in the command prompt, but you can also collect commands in a program. In MATLAB, a program or subprogram is called a *function*.

Today we are going to write a function that will take a matrix $A$ and a right-hand side $b$ as inputs and will give back a vector with the solution $Ax = b$:

```
function [x] = GaussianEliminate(A,b)
```

## 4.3 Gaussian elimination

You have probably seen Gaussian elimination before, so the following may be just a review for you.

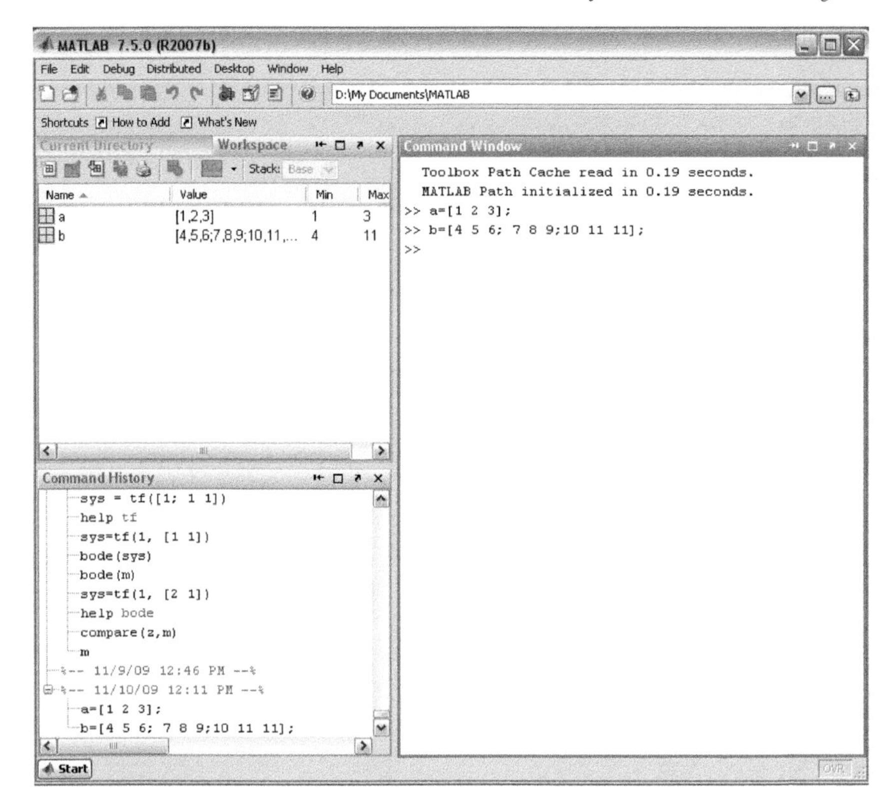


**FIGURE 4.1**
Screenshot of the MATLAB user interface with three main windows: the command prompt, the command history and the work space.

We are going to take a look at the following linear system:

$$Ax = b, \tag{4.1}$$

or

$$\begin{bmatrix} A_{11} & A_{12} & A_{31} \\ A_{21} & A_{22} & A_{23} \\ A_{31} & A_{32} & A_{33} \end{bmatrix} \begin{bmatrix} x_1 \\ x_2 \\ x_3 \end{bmatrix} = \begin{bmatrix} b_1 \\ b_2 \\ b_3 \end{bmatrix}, \tag{4.2}$$

or as an augmented matrix:

$$\left[\begin{array}{ccc|c} A_{11} & A_{21} & A_{31} & b_1 \\ A_{21} & A_{22} & A_{32} & b_2 \\ A_{31} & A_{23} & A_{33} & b_3 \end{array}\right]. \tag{4.3}$$

In MATLAB you can easily define the matrix $A$ and vector $b$ by typing (Figure 4.1):

```
>>A = [ 1 1 1; 2 1 1; 1 2 0]
>>b = [ 1 1 1];
```

In order to simplify our system, we can perform *row operations*, that is, adding multiples of equations together in order to eliminate variables.

For example, we could eliminate element $A_{21}$ by subtracting $A_{21}/A_{11} = d_{21}$ times row 1 from row 2.

In this case, row 1 is called a *pivot row* and $A_{11}$ is called the *pivot element.*

So by subtracting $d_{21}$ times row 1 from row 2, we end up with four new elements in the second row of our augmented matrix and we have eliminated element $A_{21}$ (turned into zero):

$$\begin{bmatrix} A_{11} & A_{21} & A_{31} & | & b_1 \\ 0 & A'_{22} & A'_{32} & | & b'_2 \\ A_{31} & A_{23} & A_{33} & | & b_3 \end{bmatrix}. \tag{4.4}$$

In MATLAB we could perform these operations by typing in the workspace:

```
>>d21 = A(2,1) / A(1,1)
>>A(2,1) = A(2,1) - A(1,1) * d21
>>A(2,2) = A(2,2) - A(1,2) * d21
>>A(2,3) = A(2,3) - A(1,3) * d21
>>b(2) = b(2) - b(1)*d21
```

We perform similar actions in order to eliminate element $A_{31}$, where we subtract row 1, $d_{31} = A_{31}/A_{11}$ times from row 3:

$$\begin{bmatrix} A_{11} & A_{21} & A_{31} & | & b_1 \\ 0 & A'_{22} & A'_{32} & | & b'_2 \\ 0 & A'_{23} & A'_{33} & | & b'_3 \end{bmatrix}. \tag{4.5}$$

In MATLAB we could do that by typing:

```
>>d31 = A(3,1) / A(1,1)
>>A(3,1) = A(3,1) - A(1,1) * d31
>>A(3,2) = A(3,2) - A(1,2) * d31
>>A(3,3) = A(3,3) - A(1,3) * d31
>>b(3) = b(2) - b(1)*d21
```

After having made all elements below the pivot element in the first column zero, we move to the next column. We can now use $A_{22}$ as a pivot element and eliminate $A_{33}$: subtracting $d_{31} = A_{32}/A_{22}$ times row 2 from row 3:

$$\begin{bmatrix} A_{11} & A_{21} & A_{31} & | & b_1 \\ 0 & A'_{22} & A'_{32} & | & b'_2 \\ 0 & 0 & A''_{33} & | & b''_3 \end{bmatrix}. \tag{4.6}$$

In MATLAB you would type:

```
>>d32 = A(3,2) / A(2,2)
>>A(3,2) = A(3,2) - A(2,2) * d32
>>A(3,3) = A(3,3) - A(2,3) * d32
>>b(3) = b(3) - b(1)*d32
```

Now with Equation 4.6 we have obtained a matrix of a *triangular form*, which we can easily solve with *back substitution*, because:

$$x_3 = b''_3/A''_{33} \quad (4.7)$$
$$x_2 = (b'_2 - A'_{23}x_3)/A'_{22} \quad (4.8)$$
$$x_1 = (b_1 - A_{12}x_2 - A_{13}x_3)/A_{11}. \quad (4.9)$$

Or, written in the MATLAB command prompt,

```
>>x(3) = b(3) /A(3,3)
>>x(2) = (b(2) - A(2,3)*x(3))/A(2,2)
>>x(1) = (b(1)-A(1,2)*x(2) - A(1,3)*x(3))/A(1,1)
```

Rather than typing each command into MATLAB, we could write *for loops* to automate these operations. Some tricks might come in handy, for example, we can access an entire row of a matrix by typing

```
>>A(1,:)
```

to access the first row of matrix $A$, or

```
>>A(:,2)
```

to access the second column of matrix $A$. And with

```
A(1,2:end)
```

we can access elements 2 to the last element of row 1. A row operation could look like

```
>>A(i,:) = A(i,:) - 2*A(1,:)
```

where the i-th row equals the i-th row minus two times the first row of matrix $A$.

Now we have enough information to write our program. Choose from the MATLAB menu: *file/new/m-file.* We need to make two loops, one loop that will move through the columns of the matrix and find the diagonal element on each column and eliminate each element below it, and we have to write an inner loop that will go through each row under the pivot element. Our program would look like:

```
function [x] = GaussianEliminate(A,b)
N = length(b);
for column = 1: (N-1)
 for row = (column+1):N
 d = A(row,column)/A(column,column);
 A(row,:) = A(row,:) - d(A(column,:);
 b(row) = b(row) - d*b(column);
 end
end
```

With this program we would obtain a triangular matrix that can be solved with back substitution. For each $x$ we can find its value from

$$x_i = \frac{1}{A_{i,i}} \left( b_i - \sum_{i=j+1}^{N} A_{i,j} x_j \right). \tag{4.10}$$

To include the back substitution in the program, you need to add:

```
for row=N-1:1
 x(row = b(row);
 for i = (row+1):N
 x(row)=x(row)-A(row,i)*x(i);
 end
 x(row) = x(row)/A(row,row);
end
x=x';
return
```

You can run the algorithm by typing the following at the command prompt:

```
GaussianEliminate(A,b)
```

Make sure that the current directory (the bar on the top) is set to the directory where you saved your file.

Now try to run the program with the following matrix:

$$A = \begin{bmatrix} 0 & 2 & 1 \\ 3 & 2 & 1 \\ 1 & 1 & 1 \end{bmatrix} \tag{4.11}$$

It does not work because there is a division by zero! We can easily solve the problem by *swapping* row 1 and row 2. This row swapping is called *partial pivoting*. When swapping, don't forget to also swap the right-hand side (rhs) of the equation (the $b$ vector). So, add the following code to the beginning of your program:

```
[dummy,index] = max(abs(A(column:end,column)));
index = index + column-';
temp = A(column,:);
A(column,:)=A(index,:);
A(index,:) = temp;
temp = b(column);
b(column) = b(index);
b(index) = temp;
```

Now you have written your first MATLAB program that can solve linear systems of equations. However, there are some good reasons why not to use this program. The first one is that MATLAB, itself, has a good solver to compute the solution for $Ax = b$. Our program contains many loops and it will make MATLAB slow. If you add a counter to the algorithm to monitor how many subtraction and multiplication operations are performed for a given size of matrix $A$, you will find that the number of operations for Gaussian elimination (row operations) is equal to the number of equations to the third power. For back substitution, the program requires a number of operations proportional to the square of the number of equations. Back substitution is more efficient than row operations, so maybe there are more efficient ways to end up with triangular matrices.

## 4.4 LU factorization

Suppose that we would like to solve the previous system, but with three different right-hand sides:

$$Ax_1 = b_1, Ax_2 = b_2, Ax_3 = b_3. \tag{4.12}$$

We do not really want to perform Gaussian elimination for each of the three systems, so we could write Equation 4.12 as one system:

$$A \begin{bmatrix} \vdots & \vdots & \vdots \\ x_1 & x_2 & x_3 \\ \vdots & \vdots & \vdots \end{bmatrix} = \begin{bmatrix} \vdots & \vdots & \vdots \\ b_1 & b_2 & b_3 \\ \vdots & \vdots & \vdots \end{bmatrix}. \tag{4.13}$$

By Gaussian elimination we could factor the matrix $A$ into two matrices, $L$ and $U$, so that

$$\begin{bmatrix} A_{11} & A_{12} & A_{13} \\ A_{21} & A_{22} & A_{23} \\ A_{31} & A_{32} & A_{33} \end{bmatrix} = \begin{bmatrix} 1 & 0 & 0 \\ \star & 1 & 0 \\ \star & \star & 1 \end{bmatrix} \begin{bmatrix} \star & \star & \star \\ 0 & \star & \star \\ 0 & 0 & \star \end{bmatrix}. \tag{4.14}$$

Then we could solve each right-hand side using only forward and back substitution. So as the system is now given:

$$Ax = b, \tag{4.15}$$

we could rewrite $A$ in terms of $L$ and $U$:

$$LUx = b. \tag{4.16}$$

Now, if we assume $y = Ux$, we can rewrite Equation 4.16 and solve by forward substitution as:

$$Ly = b. \tag{4.17}$$

And subsequently we solve by back substitution:

$$Ux = y. \tag{4.18}$$

So, how do we decompose $A$ as given before? When we eliminate $A_{21}$ we can keep multiplying by a matrix that undoes this operation, such that the product remains equal to $A$:

$$\begin{bmatrix} A_{11} & A_{12} & A_{13} \\ A_{21} & A_{22} & A_{23} \\ A_{31} & A_{32} & A_{33} \end{bmatrix} = \begin{bmatrix} 1 & 0 & 0 \\ d_{21} & 1 & 0 \\ 0 & 0 & 1 \end{bmatrix} \begin{bmatrix} A_{11} & A_{12} & A_{13} \\ 0 & A_{22} - d_{21}A_{12} & A_{23} - d_{21}A_{12} \\ A_{31} & A_{32} & A_{33} \end{bmatrix}. \tag{4.19}$$

The same thing, while eliminating $A_{31}$:

$$\begin{bmatrix} A_{11} & A_{12} & A_{13} \\ A_{21} & A_{22} & A_{23} \\ A_{31} & A_{32} & A_{33} \end{bmatrix} = \begin{bmatrix} 1 & 0 & 0 \\ d_{21} & 1 & 0 \\ d_{31} & 0 & 1 \end{bmatrix} \begin{bmatrix} A_{11} & A_{12} & A_{13} \\ 0 & A'_{22} & A'_{23} \\ 0 & A'_{32} & A'_{33} \end{bmatrix} \tag{4.20}$$

with $A'_{22} = A_{22} - d_{21}A_{12}$, $A'_{23} = A_{23} - d_{21}A_{12}$, $A'_{32} = A_{32} - d_{31}A_{12}$ and $A'_{33} = A_{33} - d_{31}A_{12}$. Now eliminating $A_{32}$:

$$\begin{bmatrix} A_{11} & A_{12} & A_{13} \\ A_{21} & A_{22} & A_{23} \\ A_{31} & A_{32} & A_{33} \end{bmatrix} = \begin{bmatrix} 1 & 0 & 0 \\ d_{21} & 1 & 0 \\ d_{31} & d_{32} & 1 \end{bmatrix} \begin{bmatrix} A_{11} & A_{12} & A_{13} \\ 0 & A'_{22} & A'_{23} \\ 0 & 0 & A''_{33} \end{bmatrix} \tag{4.21}$$

with $A''_{33} = A'_{33} - d_{32}A'_{23}$. Now we have completed an LU factorization, and we can solve the $L$ and $U$ matrices with forward and back substitution.

But what if we, for example, obtain a matrix like this:

$$\begin{bmatrix} A_{11} & A_{12} & A_{13} \\ A_{21} & A_{22} & A_{23} \\ A_{31} & A_{32} & A_{33} \end{bmatrix} = \begin{bmatrix} 1 & 0 & 0 \\ d_{21} & 1 & 0 \\ d_{31} & 0 & 1 \end{bmatrix} \begin{bmatrix} A_{11} & A_{12} & A_{13} \\ 0 & A'_{22} & A'_{23} \\ 0 & A'_{32} & A'_{33} \end{bmatrix}. \tag{4.22}$$

We would like to exchange rows 2 and 3. That can be done by multiplication with a *permutation matrix*, resulting in

$$\begin{bmatrix} 1 & 0 & 0 \\ 0 & 0 & 1 \\ 0 & 1 & 0 \end{bmatrix} \begin{bmatrix} A_{11} & A_{12} & A_{13} \\ A_{21} & A_{22} & A_{23} \\ A_{31} & A_{32} & A_{33} \end{bmatrix} = \begin{bmatrix} 1 & 0 & 0 \\ d_{31} & 0 & 1 \\ d_{21} & 1 & 0 \end{bmatrix} \begin{bmatrix} A_{11} & A_{12} & A_{13} \\ 0 & A'_{32} & A'_{33} \\ 0 & A'_{22} & A'_{23} \end{bmatrix}. \tag{4.23}$$

A permutation matrix is just an identity matrix whose rows have been interchanged. After the row swapping you can proceed as normal.

The general recipe for LU factorization is as follows:

1. Write down a permutation matrix.
2. Write down the matrix to decompose.

3. Promote the largest value in the column diagonal.
4. Eliminate all elements below the diagonal.
5. Move on to the next column and move the largest elements to the diagonal.
6. Eliminate the elements below the diagonal.
7. Repeat steps 5 and 6.
8. Write down $L$, $U$, and $P$.

Let's do an example: 1. Write down a permutation matrix (initially the identity matrix:

$$P = \begin{bmatrix} 1 & 0 & 0 \\ 0 & 1 & 0 \\ 0 & 0 & 1 \end{bmatrix}. \tag{4.24}$$

2. Write down the matrix you would like to decompose, for example:

$$M = \begin{bmatrix} 0 & 1 & 1 \\ 2 & 1 & 1 \\ 1 & 2 & 0 \end{bmatrix}. \tag{4.25}$$

3. Promote the largest value in the diagonal, so, starting with column 1, row swap to promote the largest value in the column to the diagonal. Do exactly the same row swap with your identity matrix $P$:

$$M = \begin{bmatrix} 2 & 1 & 1 \\ 0 & 1 & 1 \\ 1 & 2 & 0 \end{bmatrix}, P = \begin{bmatrix} 0 & 1 & 0 \\ 1 & 0 & 0 \\ 0 & 0 & 1 \end{bmatrix}. \tag{4.26}$$

4. Eliminate all elements below the diagonal, and record the multiplier $d$ that you use for elimination, so, for example, subtract 0.5 times row 1 from row 3:

$$M = \begin{bmatrix} 2 & 1 & 1 \\ 0 & 1 & 1 \\ 1 & 1.5 & -0.5 \end{bmatrix}, \begin{bmatrix} 2 & 1 & 1 \\ 0 & 1 & 1 \\ \mathbf{0.5} & 1.5 & -0.5 \end{bmatrix}. \tag{4.27}$$

5. Move to the next column. Swap rows to move the largest element to the diagonal, also for $P$:

$$\begin{bmatrix} 2 & 1 & 1 \\ \mathbf{0.5} & 1.5 & -0.5 \\ 0 & 1 & 1 \end{bmatrix}, P = \begin{bmatrix} 0 & 1 & 0 \\ 0 & 0 & 1 \\ 1 & 0 & 0 \end{bmatrix}. \tag{4.28}$$

6. Eliminate elements below the diagonal, so subtract 2/3 times row 2 from row 3:

$$\begin{bmatrix} 2 & 1 & 1 \\ 0.5 & 1.5 & -0.5 \\ 0 & 2/3 & 4/3 \end{bmatrix}. \tag{4.29}$$

7. Repeat steps 5 and 6 for all columns.
8. Write down $L$, $U$, and $P$:

$$U = \begin{bmatrix} 2 & 1 & 1 \\ 0 & 1.5 & -0.5 \\ 0 & 0 & 4/3 \end{bmatrix}, L = \begin{bmatrix} 1 & 0 & 0 \\ 0.5 & 1 & 0 \\ 0 & 2/3 & 1 \end{bmatrix}, P = U = \begin{bmatrix} 1 & 0 & 0 \\ 0 & 0 & 1 \\ 1 & 0 & 0 \end{bmatrix}. \tag{4.30}$$

In MATLAB, LU factorization can be easily done by typing:

```
>>[L,U,P] = lu(A)
```

## 4.5 Summary

In this chapter we wrote a program that can solve a system of linear equations using Gaussian elimination and back substitution. This method is rather slow for large systems. MATLAB has a good solver of `A\b` itself. We found that back substitution is relatively fast and that repeatedly performing row operations slows down the solution process a lot. Decomposing a matrix into an $L$ and a $U$ matrix can be used to perform row operations systematically and much faster. The $L$ and $U$ matrices can directly be solved using forward and back substitution. MATLAB also has a tool for LU factorizion, namely `lu`.

## 4.6 Exercises

### Exercise 1.a

Use Gaussian elimination to solve the following system:

$$\begin{aligned} 2x_1 + 3x_2 - x_3 &= 5 \\ 4x_1 + 4x_2 - 3x_3 &= 3 \\ 2x_1 - 3x_2 + x_3 &= -1. \end{aligned}$$

### Exercise 1.b

Use Gaussian elimination to solve the following system:

$$\begin{aligned} x_1 + 2x_2 + x_3 &= 3 \\ 3x_1 + 5x_2 + 2x_3 &= 1 \\ 2x_1 + 3x_2 + x_3 &= 5. \end{aligned}$$

By writing the system as $Ax = b$, you can solve it using MATLAB as follows:

```
>> x = A\b.
```

### Exercise 2.a

Solve the linear system $Ax = b$ by performing LU factorization. $A$ and $b$ are given as:

$$A = \begin{bmatrix} 1 & 1 & 1 \\ 2 & 1 & -1 \\ 4 & 1 & 5 \end{bmatrix} ; b = \begin{bmatrix} 1 \\ 2 \\ 10 \end{bmatrix}.$$

Show at every step the matrix by which you multiply the system.

### Exercise 2.b

Define the permutation matrix in terms of $L$ and $U$. You can also solve the system with MATLAB, using the following code:

```
>> A =[1, 1, 1 ; 2, 1, -1; 4, 1, 5]
>> b = [1; 2; 10];
>> [L,U,P] = lu(A)
```

Give the entries of the permutation matrix.

### Exercise 3

Assume we are solving three different linear systems with the same matrix A:

$$Ax_1 = b_1, Ax_2 = b_2, Ax_3 = b_3.$$

Run the following MATLAB code, which defines matrix A and vectors $b_i$ with random entries, computes solution vectors $x_i$, and reports the CPU time needed to compute $x_i$.

```
>> n = 300;
>> A = rand(n,n); b1 = rand(n,1); b2 = rand(n,3); b3 = rand(n,3);
>> tic; x1=A\b1; x2 = A\b2; x3 = A\b3; toc
```

Write down the reported CPU time. Now compute $x_i$ also as

```
>>tic, Ainv=inv(A); x1 = Ainv*b1; x2 = Ainv*b2; x3=Ainv*b3; toc
```

Which method is the fastest? Try to change the value of $n$ (not too small, say, larger than 100). Explain the differences in CPU time.

# 5

# Iterative methods

## 5.1 Introduction

We are going to take a look at iterative methods that can be used to solve large systems of (linear) equations.

We will solve Laplace's equation, which describes heat conduction in a rectangular geometry.

## 5.2 Laplace's equation

The equation that governs temperature in a slab of material is given by:

$$\frac{\partial T}{\partial t} = \alpha \nabla^2 T, \tag{5.1}$$

where $\alpha$ is the thermal diffusivity. $\nabla$ is the partial derivative operator. We will consider this equation as a steady-state problem, with no dependence on time:

$$\alpha \nabla^2 T = 0. \tag{5.2}$$

If we write out Equation 5.1 in two dimensions, for Cartesian coordinates we will have:

$$\frac{\partial^2 T}{\partial x^2} + \frac{\partial^2 T}{\partial y^2} = 0. \tag{5.3}$$

Figure 5.1 shows you how the domain is defined, with four boundaries. We now place a grid over the domain and we want to track the temperature at each point on the grid, as in Figure 5.2. For simplicity we will use an equally divided grid interval, so $\Delta x = \Delta y$.

We can index a node with $k = i + Nx(j-1)$ such that $T_{i,j} = T_k$. Now we can use finite differences to approximate our two-dimensional Laplace equation. Thus, we need some kind of estimate of the second derivative at node $k$.

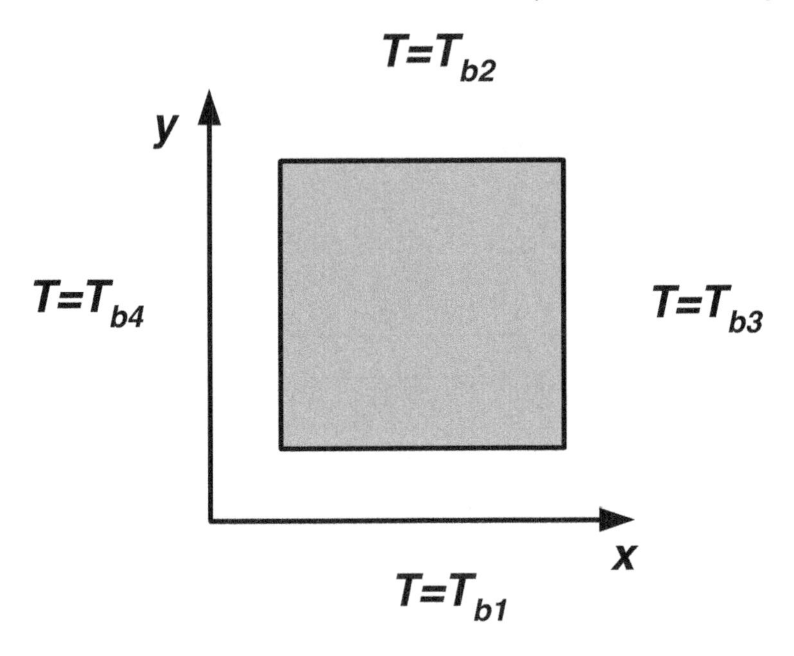


**FIGURE 5.1**
Domain definition with 4 boundaries

We could assume a piece-wise linear profile, where we approximate at each grid interval, a temperature difference by a linear function. We can estimate the second derivative by:

$$\frac{\partial^2 T}{\partial x^2} \approx = \frac{\frac{\partial T}{\partial x}\Big|_{i+1/2} - \frac{\partial T}{\partial x}\Big|_{i-1/2}}{\Delta x}, \tag{5.4}$$

which can be written as

$$\frac{\partial^2 T}{\partial x^2} \approx \frac{\frac{T_{i+1,j}-T_{i,j}}{\Delta x} - \frac{T_{i,j}-T_{i-1,j}}{\Delta x}}{\Delta x} = \frac{T_{i+1,j} + 2T_{i,j} - T_{i-1,j}}{\Delta x^2}. \tag{5.5}$$

Doing the same for the y-direction, we end up with

$$\frac{T_{i+1,j} - 2T_{i,j} + T_{i-1,j}}{\Delta x^2} + \frac{T_{i+1,j} - 2T_{i,j} + T_{i-1,j}}{\Delta y^2} = 0, \tag{5.6}$$

or, in terms of node indices, as

$$\frac{T_{k+1} - 2T_k + T_{k-1}}{\Delta x^2} + \frac{T_{k+Nx} - 2T_k + T_{k-Nx}}{\Delta y^2} = 0, \tag{5.7}$$

If we take an equally spaced grid of $\Delta x = \Delta y = 1$, we can rewrite Equation 5.7 as:

$$T_{k-Nx} + T_{k-1} - 4T_k + T_{k+1} + T_{k+Nx} = 0. \tag{5.8}$$

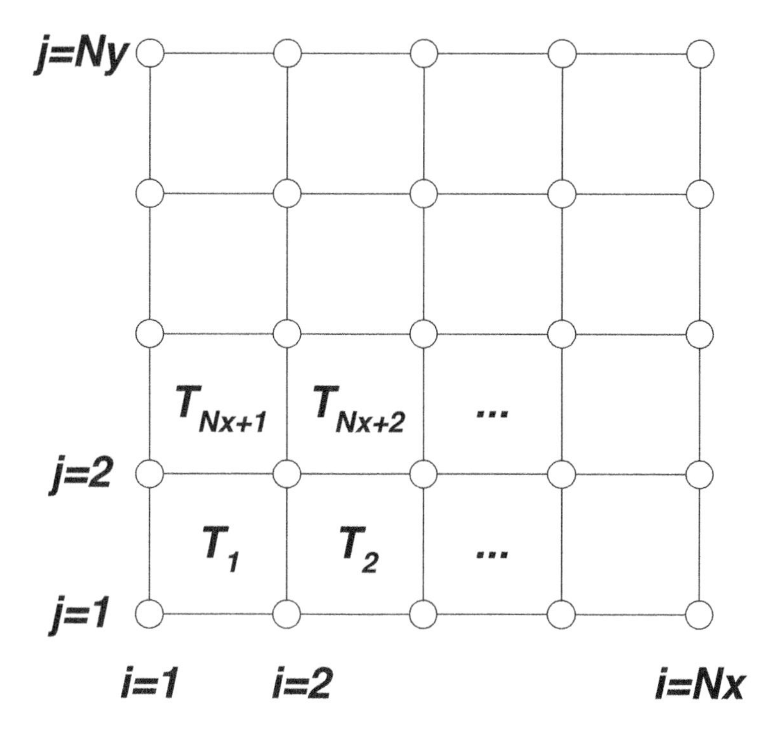


**FIGURE 5.2**
Domain definition with 4 boundaries with discretized grid

This equation tells you that the temperature at a grid point is equal to the average of the surrounding temperatures.

For the nodes on the boundaries, we have a simple equation:

$$T_{k,boundary} = \textbf{some fixed temperature}. \tag{5.9}$$

With our previous equation, and with our boundaries, we end up with a matrix equation:

$$AT = b, \tag{5.10}$$

which is a linear system. If you take $Nx = Ny = 5$, your $A$ matrix will be a 25 x 25 matrix and your $T$ and $b$ will be vectors with 25 elements.

If you want to create this system in MATLAB, you could type the code below.

```
>>Nx=5;
>>Ny=5;
>>d = 1/Nx;
>>e = ones(Nx*Ny,1);
>>A =spdiags([e,e,-4*e,e,e],[-Nx,-1,0,1,Nx],Nx*Ny,Nx*Ny);
>>A = A*alpha / d^2;
```

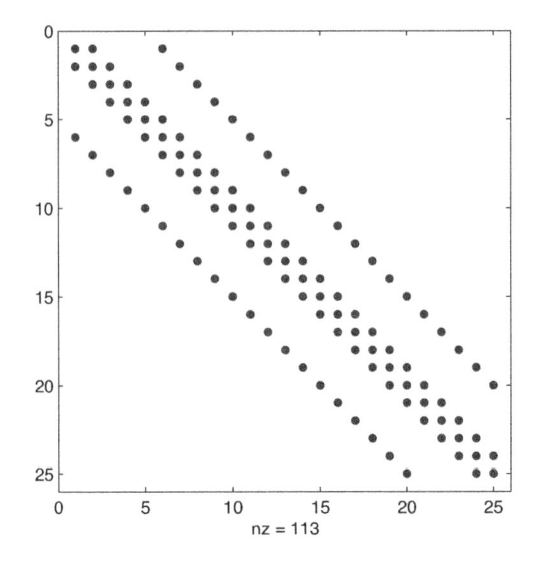


**FIGURE 5.3**
Matrix sparsity

If you "spy" (type `>>spy(A)`) the $A$ matrix, you will see that the matrix has a sparse structure, that all elements with values appear in diagonals, and that the upper and lower parts of the matrix are zeros, as shown in Figure 5.3. It is also not triagonal, as there are offset bands. Such offset bands can cause you a lot of problems! You could now solve the linear system with `T = A \b`, for the boundaries $Tb1 = 10$ and $Tb2 = Tb3 = Tb4 = 0$, and you could obtain a profile as given in Figure 5.4.

## 5.3 LU factorization

What will happen if we do Gaussian elimination by LU factorization? You could use the command `lu`. When we factorize matrix A into $L$, $U$, and $P$, we produce matrices that are less sparse than the original matrix. We have filled the elements between the offset and central band diagonals. Type the following command to do an LU factorization of $A$:

```
>>[L,U,P] = lu(A)
>>subplot(1,2,1)
>>spy(L)
```

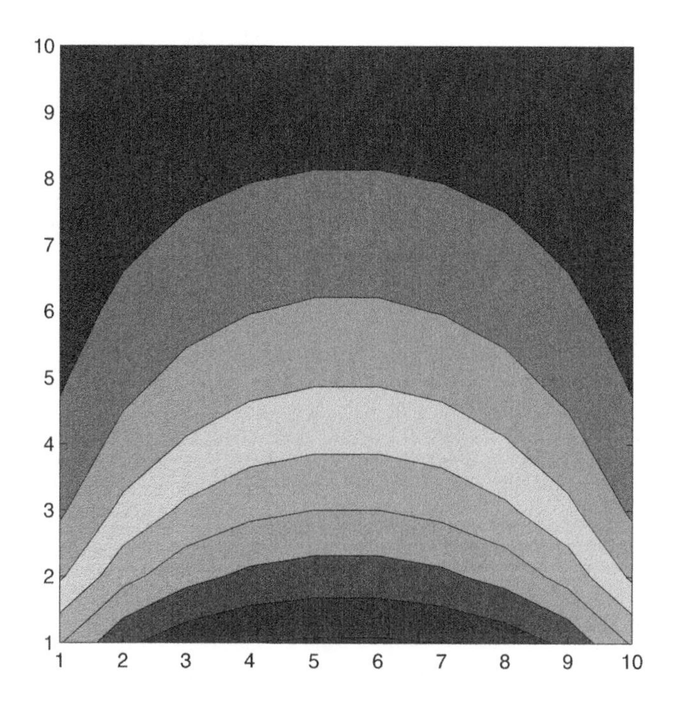


**FIGURE 5.4**
Contour plot of temperature profile

```
>>subplot(1,2,2)
>>spy(U)
```

Doing Gaussian elimination on a matrix like A requires storage of more elements as the algorithm proceeds. If we had taken a 3D problem, we would have had another offset diagonal band, even farther from the central band. The matrix produced by elimination takes up a lot of memory. But for MATLAB, this is not a problem (it allocates extra memory). MATLAB also reorders the equations so that elements are moved closer to the diagonal.

Figure 5.5 shows how the off set bands disappear after LU decomposition.

## 5.4 Iterative methods

In conclusion, Gaussian elimination is not ideal for solving sparse systems. If we are dealing with large sparse systems of equations, we can resort to iterative

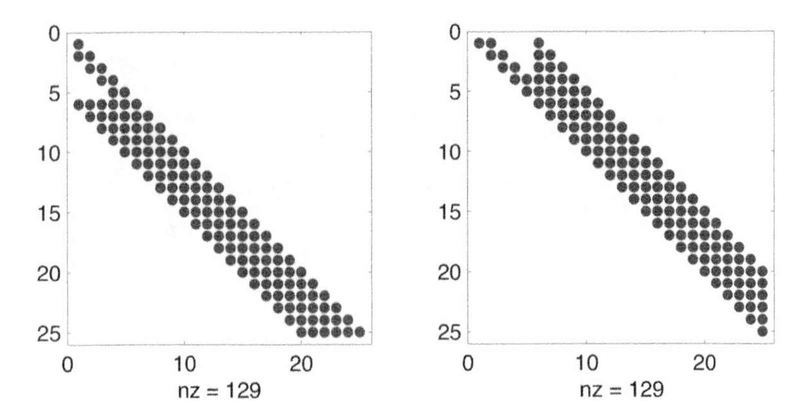


**FIGURE 5.5**
With LU decomposition, we produce matrices that are less sparse than the original matrix

methods. In the early days of computing, iterative methods were important because memory was limited.

There are several iterative methods available. For example, the Jacobi method, the Gauss-Seidel method, and successive over relaxation. MATLAB has some functions for iterative methods, and it wouldn't hurt to take a look at them (e.g., MathWorks, by typing help `bicg` or going to the MathWorks Web site). In this chapter we will look at one method, the Jacobi method.

## 5.5 The Jacobi method

As you remember, we derived an equation for steady-state heat conduction earlier:

$$T_{k-Nx} + T_{k-1} - 4T_k + T_{k+1} + T_{k+Nx} = 0. \tag{5.11}$$

We could rearrange that equation into:

$$T_k = \frac{T_{k-Nx} + T_{k-1} + T_{k+1} + T_{k+Nx}}{4}. \tag{5.12}$$

In the Jacobi scheme, the iteration continues with an initial guess for the values of $T$ at each node, and we make a new updated value using Equation 5.12 and obtaining Equation 5.15:

$$T_{k,new} = \frac{T_{k-Nx,old} + T_{k-1,old} + T_{k+1,old} + T_{k+Nx,old}}{4}. \tag{5.13}$$

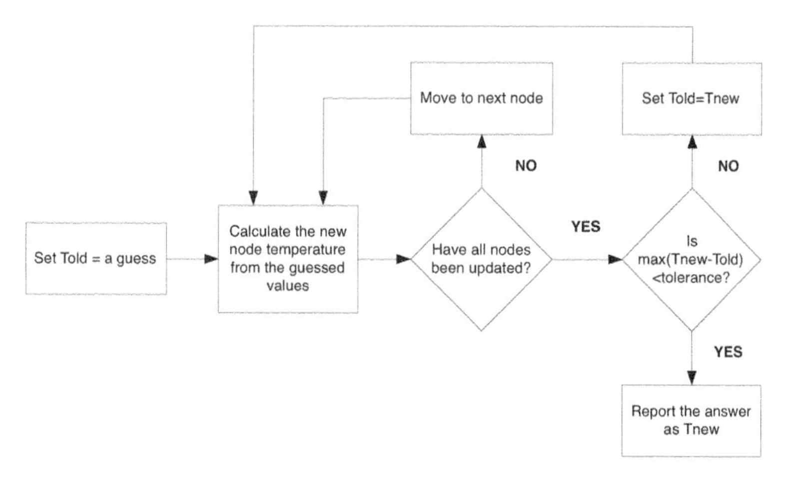


**FIGURE 5.6**
Jacobi scheme where only two vectors need to be stored in this iterative procedure $T_{old}$ and $T_{new}$

We do this for all other nodes. Figure 5.6 shows the algorithm for executing the Jacobi scheme in solving the Laplace equation. Once we have gone through all nodes, we use the new guess values as a guess and repeat. We could use the approach, using matrices, where we start with a matrix $A$, generally of the following form:

$$A = \begin{bmatrix} * & * & & * & & \\ * & * & * & & * & \\ & * & * & * & & * \\ * & & * & * & * & \\ & * & & * & * & * \\ & & * & & * & * \end{bmatrix}. \tag{5.14}$$

We split this matrix into a diagonal matrix $D$:

$$D = \begin{bmatrix} * & & & & & \\ & * & & & & \\ & & * & & & \\ & & & * & & \\ & & & & * & \\ & & & & & * \end{bmatrix}, \tag{5.15}$$

and another matrix $S$:

$$S = \begin{bmatrix} & * & & * & & \\ * & & * & & * & \\ & * & & * & & * \\ * & & * & & * & \\ & * & & * & & * \\ & & * & & * & \end{bmatrix}, \tag{5.16}$$

so

$$A = D + S \tag{5.17}$$

We can solve $AT = b$ by the following steps:

$$(D + S)T = b, \tag{5.18}$$

rewriting:

$$DT = b - ST. \tag{5.19}$$

Suppose we want to calculate a new value. We could state:

$$DT_{new} = b - ST_{old}, \tag{5.20}$$

or, in other words:

$$T_{new} = D^{-1}(b - ST_{old}). \tag{5.21}$$

If we define an error at the $k$-th iteration:

$$\varepsilon_k = T_k - T_{k-1} \tag{5.22}$$

and substitute this into the previous equation, we obtain a formulation of the error at the next iteration:

$$T_{k+1} = D^{-1}(b - ST_k). \tag{5.23}$$

So:

$$\varepsilon_{k+1}T = D^{-1}(b - S\varepsilon_k - ST). \tag{5.24}$$

In other words:

$$D\varepsilon_{k+1} = -S\varepsilon_k, \tag{5.25}$$

which can be rewritten as:

$$\varepsilon_{k+1} = -D^{-1}S\varepsilon_k. \tag{5.26}$$

The eigenvalues of $D^{-1}S$ must always have a modulus less than unity to ensure that the error of the new iteration is smaller than the previous one (otherwise the error propagates!).

We could express the error vector in terms of eigenvectors of $D^{-1}S$ as

$$\varepsilon = a_1u_1 + a_2u_2 + a_3u_3 + a_4u_4 + \cdots, \tag{5.27}$$

so the error at the $k + 1$-th position would be like

$$\varepsilon_{k+1} = D^{-1}S(a_1u_1 + a_2u_2 + a_3u_3 + a_4u_4 + \cdots), \tag{5.28}$$

which is

$$\varepsilon_{k+1} = a_1D^{-1}Su_1 + a_2D^{-1}Su_2 + a_3D^{-1}Su_3 + a_4D^{-1}Su_4 + \cdots, \tag{5.29}$$

which is

$$\varepsilon_{k+1} = a_1\lambda_1 u_1 + a_2\lambda_2 u_2 + a_3\lambda_3 u_3 + a_4\lambda_4 u_4 + \cdots \tag{5.30}$$

The magnitude of the error will grow on each iteration if any of the eigenvalues ($\lambda$) have a complex modulus greater than unity! We should find the largest magnitude of eigenvalue for the matrix $D^{-1}S$.

The eigenvalues must be within this circle for the method to converge. To estimate the eigenvalues, we can use Gershgorin's theorem, which states:

*For a square matrix $M$ and row $k$, an eigenvalue is located on the complex plane within a radius equal to or less than the sum of the moduli of the off-diagonal elements of that row.*

This can be formulated in an equation:

$$|\lambda - m_{k,k}| \leq \sum_{j=1, j\neq k}^{N} |m_{k,j}| \, . \tag{5.31}$$

That means, for the Jacobi iteration, the off-diagonal elements of row $k$ of $D^{-1}S$ are $1/a_{k,k}$ times the off-diagonal elements of the original matrix $A$, while the diagonal element is zero. Combining Gershgorin's theorem with $\lambda \leq 1$ and the structure of the matrix $D^{-1}S$, we can derive

$$|a_{k,k}| \geq \sum_{j=1, j\neq k}^{N} |a_{k,j}| \, . \tag{5.32}$$

In other words, for the Jacobi method to be stable, the size of the diagonal element must be larger than the sum of the moduli of the other elements in the row. Such a matrix is called *diagonally dominant.* It turns out that for the sorts of banded matrices you find when solving PDEs, this condition is required for stability.

## 5.6 Example for the Jacobi method

Let us work out an example to get some feel for the Jacobi method. Suppose we want to find a solution to the following linear system:

$$\begin{bmatrix} 2 & 0 & 1 \\ 0 & 2 & -1 \\ -1 & -1 & 4 \end{bmatrix} \begin{bmatrix} x_1 \\ x_2 \\ x_3 \end{bmatrix} = \begin{bmatrix} 1 \\ 2 \\ -1 \end{bmatrix} . \tag{5.33}$$

We are going to write our $M$ matrix as the sum of the diagonal matrix $D$ and another matrix $S$. That's easy:

$$D = \begin{bmatrix} 2 & 0 & 0 \\ 0 & 2 & 0 \\ 0 & 0 & 4 \end{bmatrix}. \tag{5.34}$$

And

$$S = \begin{bmatrix} 0 & 0 & 1 \\ 0 & 0 & -1 \\ 1 & -1 & 0 \end{bmatrix}. \tag{5.35}$$

For our calculations we also need the inverse of $D$. For diagonal matrices, the inverse can be calculated easily:

$$D^{-1} = \begin{bmatrix} 1/d_{1,1} & 0 & 0 \\ 0 & 1/d_{2,2} & 0 \\ 0 & 0 & 1/d_{3,3} \end{bmatrix}. \tag{5.36}$$

So, for our example, the inverse of $D$ is

$$D^{-1} = \begin{bmatrix} 0.5 & 0 & 0 \\ 0 & 0.5 & 0 \\ 0 & 0 & 0.25 \end{bmatrix}. \tag{5.37}$$

We have to start with an initial guess for our $x$ values, e.g., $x_0 = [0, 0, 0]$, and we have to define a solution tolerance $\varepsilon \leq 0.1$.

Recall that the Jacobi scheme was given as

$$x_{new} = D^{-1}(b - Sx_{old}). \tag{5.38}$$

For our example we will find:

$$x_1 = \begin{bmatrix} 0.5 & 0 & 0 \\ 0 & 0.5 & 0 \\ 0 & 0 & 0.25 \end{bmatrix} \left( \begin{bmatrix} 1 \\ 2 \\ -1 \end{bmatrix} - \begin{bmatrix} 0 & 0 & 1 \\ 0 & 0 & -1 \\ 1 & -1 & 0 \end{bmatrix} \begin{bmatrix} 0 \\ 0 \\ 0 \end{bmatrix} \right) = \begin{bmatrix} 0.5 \\ 1 \\ -0.25 \end{bmatrix}. \tag{5.39}$$

If we use the two-norm as tolerance criterion, we will find for $\varepsilon$ after the first iteration:

$$||x_1 - x_0||_2 = 1.1456. \tag{5.40}$$

We now use $x_1$ as the values to calculate a new approximation of the solution $x_2$ exactly the same way:

$$x_2 = \begin{bmatrix} 0.5 & 0 & 0 \\ 0 & 0.5 & 0 \\ 0 & 0 & 0.25 \end{bmatrix} \left( \begin{bmatrix} 1 \\ 2 \\ -1 \end{bmatrix} - \begin{bmatrix} 0 & 0 & 1 \\ 0 & 0 & -1 \\ 1 & -1 & 0 \end{bmatrix} \begin{bmatrix} 0.5 \\ 1 \\ -0.25 \end{bmatrix} \right) = \begin{bmatrix} 0.625 \\ 0.875 \\ -0.125 \end{bmatrix}. \tag{5.41}$$

And the two-norm is:

$$||x_2 - x_1||_2 = 0.2165. \tag{5.42}$$

At the third iteration you will find $x_3 = [0.5625, 0.9375, -0.1875]$ with $||x_3 - x_2||_2 = 0.10825$. And at the fourth iteration you have already converged: $x_4 = [0.59375, 0.90625, -0.15625]$ with $||x_4 - x_3||_2 = 0.0541$.

## 5.7 Summary

The main message of this chapter was that PDEs can be written as sparse systems of linear equations. You could use a direct method, like Gaussian elimination to solve such system, but if you have systems in more than one dimension, there are more efficient alternatives such as iterative methods. You saw, briefly, how the Jacobi method works, and you found that eigenvalues of composite matrices, representing error in the iteration, determine whether the method is stable or not (that means the error converges).

## 5.8 Exercises

### Exercise 1

Given is the Laplace equation

$$\frac{\partial^2 T}{\partial x^2} + \frac{\partial^2 T}{\partial y^2} = 0, \tag{5.43}$$

where the domain is bounded on a rectangular grid by four constant temperature boundaries: $T_{b1} = 10$ and $T_{b2} = T_{b3} = T_{b4} = 0$. The initial condition that holds is $T_0 = 0$.

Make a discrete approximation of Equation 15.8 using finite differences.

### Exercise 2

We can write the discretized equations formulated in Example 1 as a matrix equation system:

$$A\mathbf{T} = b. \tag{5.44}$$

If we take 5 points along the x axis and 5 points along the y axis, how would matrix $A$ and vector $b$ look?

Type the following code to obtain $A$:

```
>> Ny = 5; Nx = 5; d = 1/Nx; alpha = 1;
>> e = ones(Nx*Ny,1);
>> A = spdiags([e,e,-4*e,e,e],[-Nx,-1,0,1,Nx],Nx*Ny,Nx*Ny);
>> A = A*alpha/d^2;
```

and $b$:

```
b = linspace(0,0,Nx*Ny)';
for i = 1:Nx
 j = 1;
 k = i + (Nx)*(j-1);
 b(k) = b(k) + Tb1*alpha / d^2;
 j = Ny;
 k = i + (Nx)*(j-1);
 b(k) = b(k) + Tb2*alpha / d^2;
end
for j = 1:Ny
 i = 1;
 k = i + (Nx)*(j-1);
 b(k) = b(k) + Tb3*alpha / d^2;
 if (k-1>0)
 A(k,k-1) = 0;
 end
 i = Nx;
 k = i + (Nx)*(j-1);
 b(k) = b(k) + Tb4*alpha/ d^2;
 if (k+1<Nx*Ny)
 A(k,k+1) = 0;
 end
end
```

You can check the sparsity of matrix $A$ with the command `spy(A)`.

### Exercise 3

Perform LU decomposition on $A$ and again check the sparsity of the $L$ and $U$ matrices. What do you see?

Here is how you decompose $A$ into $L$ and $U$:

```
>>[L,U,P] = lu(A)
```

# 6

# *Nonlinear equations*

## 6.1 Introduction

In this chapter we will develop a program to solve nonlinear equations using Newton's method. Before we can start with that, we need to get familiar with something from programming called a "function handle."

MATLAB can pass function names as arguments to functions. This mechanism is called a *function handle*. For example, if we want to solve the function $x^2 - 2x = 0$, we need to write a function that returns the value of $x^2 - 2x$. We could do that by typing the following code in the script editor:

```
Function [residual] = myfunc(x)
residual = x^2 -2*x
Return
```

Now we could, for example, use the routine `fzero` to solve the function and find the roots by `>>fzero(@myfunc,2)`. `@myfunc` is called a handle to the function `myfun`. In this case the number 2 is our initial guess.

## 6.2 Newton method 1D

Now, if we want to solve $x^2 - 2x = 0$, we could use Newton's method. This is something you probably did in high school. We start with an initial guess $x_0$ and calculate $f_{new}$:

$$f_{new} = f_0 + \left(\frac{\partial f}{\partial x}\right)_{x_0} (x_{new} - x_0), \tag{6.1}$$

then we extrapolate (assuming $f(x)$ is linear) to a new value of $x$ which will make $f(x) = 0$

$$(x_{new} - x_0) = \Delta x = \frac{-f}{\left(\frac{\partial f}{\partial x}\right)_{x_0}}. \tag{6.2}$$

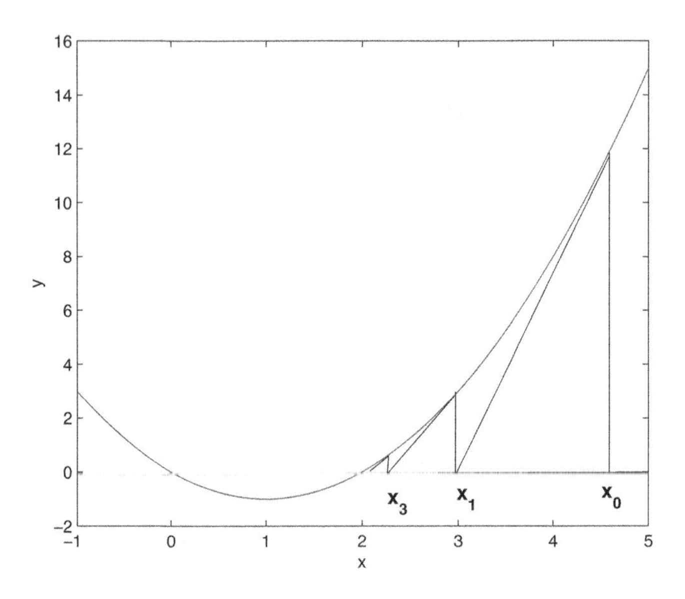


**FIGURE 6.1**
Graphical Newton method

This procedure is repeated until the solution is reached. See Figure 6.1 on how Newton's method looks graphically. The program is very simple in MATLAB. You have to define a function handle for the function you want to solve (`myfunc`). You have to supply a function for the gradient (`gradient`), and an initial guess and a tolerance.

```
function [solution] = Newton1D(myfunc,gradient,guess,tol)
x = guess;
error = 2*tol
while error > tol
 F = feval(myfunc,x);
 G = feval(gradient,x);
 dx = (-F / G);
 x = x + dx;
 F = feval(myfunc,x);
 error = (abs(F));
end
solution = x;
return
```

## 6.3 Newton method 2D

We could also use Newton's method for two or even more equations. Consider the system:

$$f_1 = x_1^3 + x_2^2 = 0 \tag{6.3}$$
$$f_2 = x_1^2 + x_2^3 = 0, \tag{6.4}$$

or in vector notation,

$$F\begin{pmatrix} x_1 \\ x_2 \end{pmatrix} = \begin{bmatrix} f_1 \\ f_2 \end{bmatrix} = \begin{bmatrix} x_1^3 + x_2^2 \\ x_1^2 + x_2^3 \end{bmatrix}. \tag{6.5}$$

We can start with a guessed value, $x_i$, and find a better guess by extrapolation. Thus, near the guessed value, the function can be expanded as

$$\begin{bmatrix} f_1(x_i + dx) \\ f_2(x_i + dx) \\ \vdots \end{bmatrix} = \begin{bmatrix} f_1(x_i) \\ f_2(x_i) \\ \vdots \end{bmatrix} + \begin{bmatrix} \frac{\partial f_1}{\partial x_1}dx_1 & \frac{\partial f_1}{\partial x_2}dx_2 & \frac{\partial f_1}{\partial x_3}dx_3 & \cdots \\ \frac{\partial f_2}{\partial x_1}dx_1 & \frac{\partial f_2}{\partial x_2}dx_2 & \frac{\partial f_2}{\partial x_3}dx_3 & \cdots \\ \vdots & \vdots & \vdots & \ddots \end{bmatrix} \tag{6.6}$$

or, if we take out the $dx$s, we obtain

$$\begin{bmatrix} f_1(x_i + dx) \\ f_2(x_i + dx) \\ \vdots \end{bmatrix} = \begin{bmatrix} f_1(x_i) \\ f_2(x_i) \\ \vdots \end{bmatrix} + \begin{bmatrix} \frac{\partial f_1}{\partial x_1} & \frac{\partial f_1}{\partial x_2} & \frac{\partial f_1}{\partial x_3} & \cdots \\ \frac{\partial f_2}{\partial x_1} & \frac{\partial f_2}{\partial x_2} & \frac{\partial f_2}{\partial x_3} & \cdots \\ \vdots & \vdots & \vdots & \ddots \end{bmatrix} \begin{bmatrix} dx_1 \\ dx_2 \\ \vdots \end{bmatrix} \tag{6.7}$$

or in short form:

$$F(x_i + dx) = F(x_i) + Jdx. \tag{6.8}$$

The matrix with the derivatives is called the Jacobian. Do you see the similarities with the 1D Newton method?

The multi-dimensional Newton method can be applied in a procedure of five steps:

- Compute the function values at the guessed value of $x_i = [x_1, x_2, x_3]^T$.
- Calculate the Jacobian matrix using the current guess.
- Solve the linear system $-F(x_i) = Jdx$ for the values of $dx$.
- Update the guessed value $x_{i+1} = x_i + dx$.
- Repeat until the updated value of $x_i$ gives an $F(x)$ sufficiently close to zero.

You could define the error criterion as the maximum value in the residual vector:

$$\max\left(f_1(x), f_2(x), \cdots\right) \leq \varepsilon \tag{6.9}$$

or, you could use the two-norm, which is fancier:

$$||F(x)||_2 = \left(\sum_i f_i^2\right)^{1/2}. \tag{6.10}$$

The MATLAB code for the multi-dimensional Newton method is as follows:

```
function [solution] = Newton1D(myfunc,jacobian,guess,tol)
x = guess;
error = 2*tol
while error > tol
 F = feval(myfunc,x);
 J = feval(jacobian,x);
 dx = -J \(-F);
 x = x + dx;
 F = feval(myfunc,x);
 error = max(abs(F));
end
solution = x;
return
```

The differences with the 1D method are very small.

As an illustration for our two-dimensional example, Figure 6.2 shows the two-norm of the system. In Figure 6.3 a contour plot is given of the same problem. We could run our Newton routine and plot the calculated coordinate pair at each iteration, which gives a plot like the one in Figure 6.3. If your start guess is far away from the solution, the trajectory that the Newton method follows to find the solution can be very 'erratic'.

## 6.4 Reduced Newton step method

One possible method to deal with potential erratic behavior and increase the efficiency is by the reduced Newton step method; see Figure 6.4. We search along the direction of the Newton step for a point where the error is less than the error at the starting point. This method is much more robust. Figure 6.4 shows the scheme for the reduced step method. There are some problems with Newton's method. When the Jacobian is singular, we cannot find a solution to $-F(x_i) = J(x_i + 1 - x_i)$, so the vector $dx$ cannot be updated. The same thing happens for the 1D method, when a maximum or minimum in the function is encountered (zero gradient).

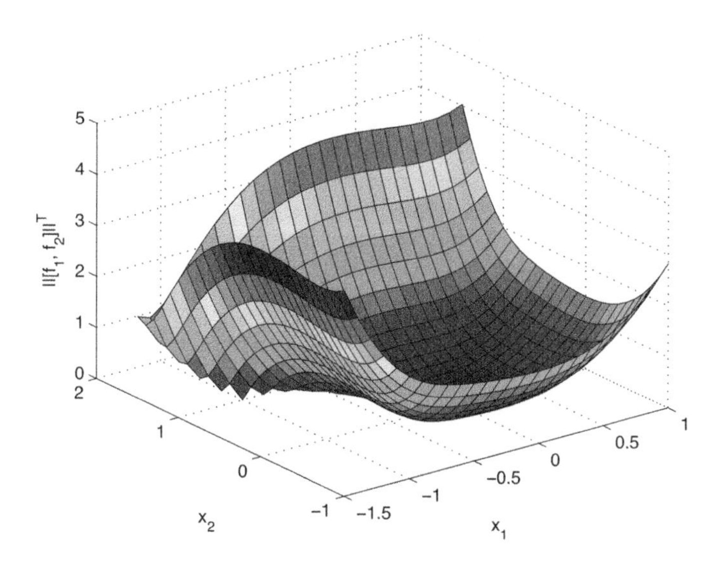


**FIGURE 6.2**
Two-norm of the 2D example

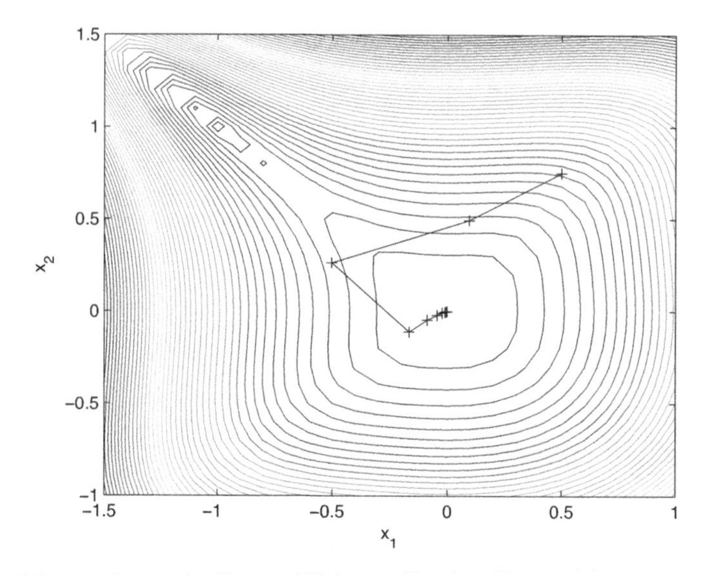


**FIGURE 6.3**
Contour plot of the two-norm of the 2D example, with the trajectory that Newton's method follows to find the solution

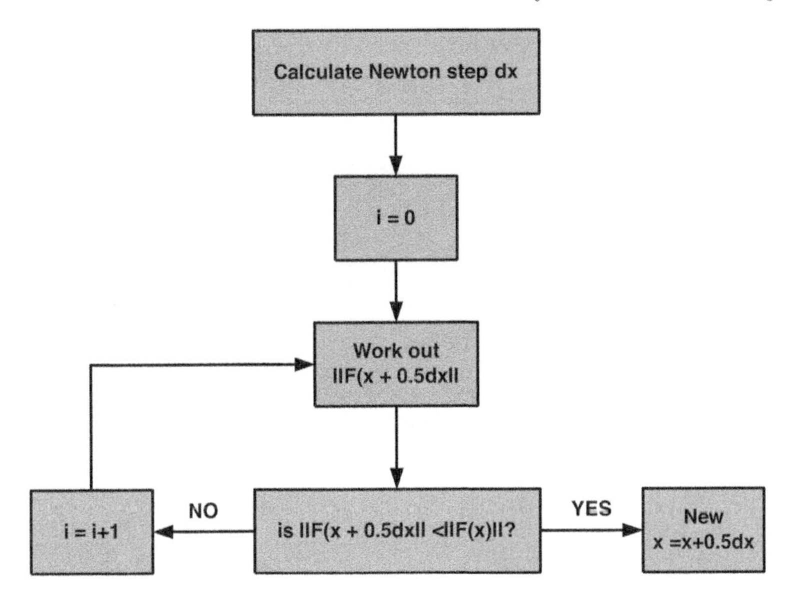


**FIGURE 6.4**
Scheme for the reduced Newton step method

The error for a given iteration can be given as the difference between the value at the iteration and the true solution:

$$\varepsilon_i = x_i - x^*. \tag{6.11}$$

Assuming that we are close to the solution, we could expand the function $F$ around the solution

$$F(x) = F^* + J^*(x - x^*) + O[(x - x^*)]^2 \tag{6.12}$$

from which follows

$$F(x) = J^*(x - x^*) + O[(x - x^*)]^2 \tag{6.13}$$

$$F(x) = J^*\varepsilon + O(\varepsilon^2). \tag{6.14}$$

Now we substitute Equations 6.11, 6.12, and 6.4 into $-F(x_i) = J(x_{i+1} - x_i)$ and obtain:

$$-J^*(\varepsilon_i) - O[(\varepsilon^2)] = J(x_{i+1} - x_i) \tag{6.15}$$

$$-J^*(\varepsilon_i) - O[(\varepsilon^2)] = J(\varepsilon_{i+1} - \varepsilon_i), \tag{6.16}$$

from which it can be concluded that the convergence of the method is quadratic (that is not very good).

We assumed that we can have an analytical form of the Jacobian, but often, when we have large systems, the functions are too expensive to evaluate.

Another way of approximating the Jacobian is by means of finite differences, where a derivative is given as

$$\frac{\partial f_i}{\partial x_i} = \frac{f_i(x) + f_i(x+\delta)}{\delta}. \tag{6.17}$$

However, $\delta$ cannot be made too small, because computers are only able to store numbers at finite precision. A common value for $\delta$ is the square root of eps (eps $= 2.2204 * 10^{-16}$). However, calculating the Jacobian is less efficient than the analytical form.

## 6.5 Quasi-Newton method

You do not need to calculate the exact value of the Jacobian at each iteration. You could use a Quasi-Newton method. A famous example is Broyden's method.

Broyden's method can be used to estimate the Jacobian. The residual is given as

$$F_{i+1}(x) \approx F_i(x) + J_i(x_{i+1} - x_i). \tag{6.18}$$

If we replace the Jacobian by an estimate we will obtain

$$F_i(x) \approx F_{i+1}(x) + B_{i+1}(x_i - x_{i+1}). \tag{6.19}$$

If we use $B_i$ (the estimate of the Jacobian at iteration i), we get

$$F_i(x) = B_i(x_{i+1} - x_i) = B_i(\Delta x_i). \tag{6.20}$$

If we now combine Equation 6.19 and Equation 6.20 we will get

$$-B_i(x_{i+1} - x_i) \approx F_{i+1}(x) + B_{i+1}(x_i - x_{i+1}) \tag{6.21}$$

$$-B_i \Delta x_i \approx F_{i+1} - B_{i+1}\Delta x_i. \tag{6.22}$$

Multiplying both sides by $(\Delta x_i)^T$ will give

$$B_{i+1} = B_i + \frac{F_{i+1}(x)(\Delta x_i)^T}{||\Delta x_i||^2}. \tag{6.23}$$

On the i-th iteration we calculate an updated estimate of the Jacobian from information calculated in the previous iteration. This kind of updating results in a linear converge, often called a quasi-Newton scheme.

Frequently a set of equations that we want to solve will be sparse (the Jacobian will contain mainly zeros). We do not need to calculate these entries. We can make use of the MATLAB operators \ and `fsolve` to deal in a handy way with sparsity!

## 6.6 Summary

To solve nonlinear equations we need an iterative procedure, for example, Newton's method. We found out how we can solve a 1D problem, and a multidimensional problem, using Newton's method. If the initial guess is far from the solution, Newton's method can be erratic. For this reason, you could employ a reduced gradient step, the steps become smaller, and the trajectory to the solution becomes straighter. We also found out that Newton's method does not work when the Jacobian is singular and that the error has a quadratic convergence with the Jacobian. Often we cannot have analytical representations of the Jacobian, and we can use finite differences to approximate the Jacobian. This is, however, inefficient. On the basis of a calculation of the Jacobian at a certain iteration, we can estimate a new value for the Jacobian at the new iteration. The method in which we update estimates for the Jacobian is called Broyden's method. Using Broyden's method, the convergence of error becomes linear or quasi-Newtonian.

## 6.7 Exercises

### Exercise 1

In this exercise we are going to write the code to solve the two-dimensional problem:

$$f_1 = x_1^3 + x_2^2 = 0 \tag{6.24}$$

$$f_2 = x_1^2 - x_2^3 = 0. \tag{6.25}$$

Write the function Newton.m, that solves the problem as:

```
function [solution] = Newton(MyFunc,Jacobian,Guess,tol)
x = Guess;
error = 2*tol
while error> tol
 F = feval(Myfunc,x);
 J = feval(Jacobian,x);
 dx = J\(-F);
 x = x+dx;
 F = feval(MyFunc,x);
 error = max(abs(F));
end
solution = x;
return
```

### Exercise 2

Now we need to write a function that contains the nonlinear system, and a function that determines the Jacobian:

```
function y = Func(x)
y(1) = x(1).^3 + x(2).^2; y(2) = x(1).^2 - x(2).^3;
y = y';
return
```

and

```
function J = Jac(x)
J(1,1) = 3*x(1) ^2;
J(1,2) = 2*x(2);
J(2,1) = 2*x(1);
J(2,2) = -4*x(2) ^2;
return
```

Save the functions, and solve the system with the following:

```
solution = Newton(@Func,@Jac,[1;1],1e-6);
```

### Exercise 3

A good measure for the residual error is the two-norm given as

$$||F(x)|| = \left(\sum_i f_i^2\right)^{1/2}. \tag{6.26}$$

Try to write a routine in MATLAB that calculates the two-norm based on the nonlinear system of Exercise 1. Then make a surface plot (using `surf(x,y,z)`. Subsequently, make a contour plot using `contour(x,y,z,N)` ($N$ is the number of contour lines that you want). Hint: you can use two for-loops to calculate the two-norm for all combinations of $x_1$ and $x_2$.

In order to see the trajectory that the Newton routine followed to find the solution, you could plot, within the contour plot, the path that the Newton routine used.

### Exercise 4

The Underwood equation for multicomponent distillation is given as

$$\left(\sum_{j=1}^{n} \frac{\alpha_j z_{jF} F}{\alpha_j - \phi}\right) - F(1-q) = 0, \tag{6.27}$$

where $F$ is the molar feed flow rate, $n$ is the number of components in the feed, $z_{jF}$ is the mole fraction of each component in the feed, $q$ is the quality of the feed, $\alpha_j$ is the relative volatility of each component at average column conditions, and $\phi$ is the root of the equation.

It has been shown by Underwood that $(n-1)$ of the roots of this equation lie between the values of the relative volatilities as shown below:

$$\alpha_n < \phi_{n-1} < \alpha_{n-1} < \phi_{n-2} < ... < \alpha_3 < \phi_2 < \alpha_2 < \phi_1 < \alpha_1. \tag{6.28}$$

Evaluate the $(n-1)$ roots of this equations for the case shown in the table below

| Component in feed | Mole fraction, $z_{jF}$ | Relative volatility, $\alpha_j$ |
|---|---|---|
| 1 | 0.05 | 10.00 |
| 2 | 0.05 | 5.00 |
| 3 | 0.10 | 2.05 |
| 4 | 0.30 | 2.00 |
| 5 | 0.05 | 1.50 |
| 6 | 0.30 | 1.00 |
| 7 | 0.10 | 0.90 |
| 8 | 0.05 | 0.10 |

where $F = 100mol.h^{-1}$ and $q = 1.0$ (saturated liquid).

# 7

# *Ordinary differential equations*

## 7.1 Introduction

In this chapter we are going to solve a common engineering problem: the initial value problem. Besides the formulation of a solution methodology, we will evaluate stability of the solution method.

## 7.2 Euler's method

The Euler method is a simple way of solving a differential equation. If the following ODE (ordinary differential equation) is given:

$$\frac{dx}{dt} = f(x,t) \tag{7.1}$$

with the initial condition $x(t = 0) = x_0$, we could generate an estimate of $x$ at $t + \delta t$ as

$$x(t + \delta t) = x(t) + \frac{dx}{dt}\delta t = x(t) + f(x,t)\delta t, \tag{7.2}$$

so we can step forward in time, by evaluating the gradient, from the current step to the next step. Figure 7.1 shows how this looks graphically.

If you wrote Euler's method in a MATLAB script, it would look like the example below, a function that has as input a function handle, an initial value, the time domain, and the number of steps you want to take.

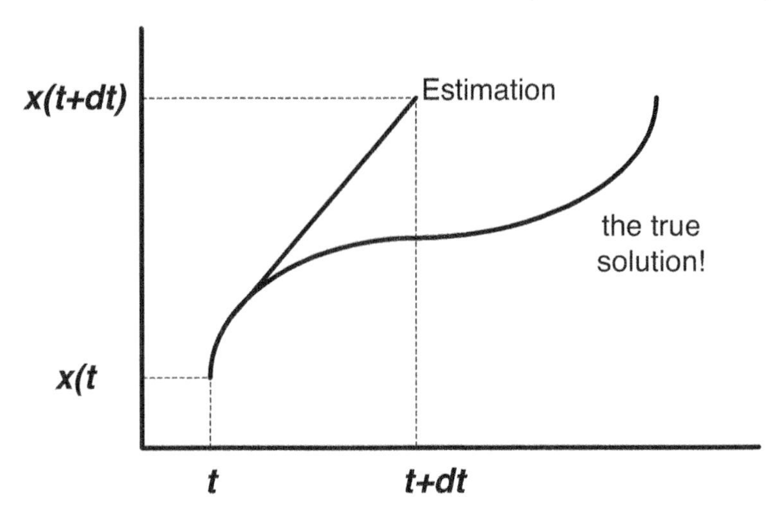


**FIGURE 7.1**
With Euler's method we can step forward in time

```
function[x,t] = Euler(MyFunc,InitialValue,Start,Finish,Nsteps)
x(1) = InitialValue;
t(1) = Start;
dt = (Finish-Start)/Nsteps;
for i =1:Nsteps
 F = feval(MyFunc,x(i),t(i));
 t(i+1) = dt +t(i);
 x(i+1) = F*dt + x(i);
end
t = t'; x = x';
return
```

We can test the routine on a simple problem, say a batch reactor, in which the reactant is consumed by a first-order reaction:

$$\frac{dx}{dt} = -kx, \tag{7.3}$$

given the initial concentration $x(t = 0) = 1$. The analytical solution will give you an exponential decay with a time constant $\tau = 1/k$.

You need to supply a function handle with the equation:

```
function [f] = TestFunction(x,t)
f = -1*x;
return
```

And now, using the Euler method, you will find the graphical result in Figure 7.2. Of course, this equation is so simple that you can also calculate it by

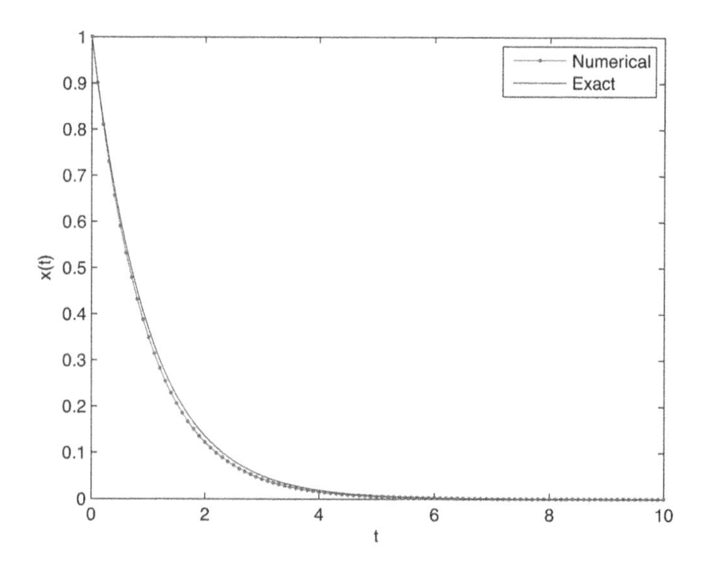


**FIGURE 7.2**
For 100 steps, the numerical solution gives a good match

hand and plot the real solution'. As you can see, the numerical solution gives a good match to the exact solution. Now, if you only take 25 steps, you see that the numerical solution is still OK, but it starts deviating a little from the exact solution; see Figure 7.3.

For 6 steps, the numerical solution follows the trend, but it already starts oscillating, while for 5 steps the numerical solution becomes unstable. If you further decrease the number of steps to, for example, only 3 steps, the numerical solution moves away from the actual solution.

## 7.3 Accuracy and stability of Euler's method

The error for a single step of the Euler method can be found from a Taylor series expansion:

$$x(t + \delta t) = x(t) + \frac{dx}{dt}\delta t = x(t) + f(x, t)\delta t, \tag{7.4}$$

in which the terms $O(\delta t^2)$ are neglected. We may assume that the error for one step will be equal to the additional terms $O(\delta t^2)$. Now we can formulate

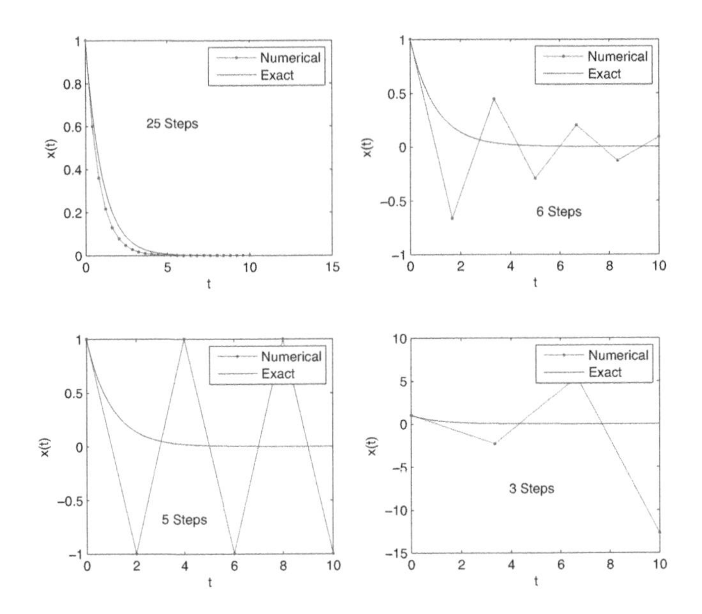


**FIGURE 7.3**
Numerical solutions for 25, 6, 5 and 3 steps

an estimate of the accumulated error for all steps, as:

$$NO(\delta t^2) = \frac{\Delta t}{\delta t} O(\delta t^2). \tag{7.5}$$

The Euler method is only accurate to $O(\delta t)$, which is not very impressive. If we take a look at the simple first-order ODE

$$\frac{dx}{dt} = \lambda x, \tag{7.6}$$

with $x(t = 0) = 1$ we can find an exact solution:

$$x = \exp(\lambda t). \tag{7.7}$$

For $\lambda < 0$, the solution decays to zero, so a numerical solution should do the same.

If we write out the Euler scheme:

$$\begin{aligned} x_{i+1}(t + \delta t) &= x_i(t) + f(x_i)\delta t && (7.8)\\ &= x_i + \lambda x_i \delta t && (7.9)\\ &= x_i(1 + \lambda \delta t) && (7.10) \end{aligned}$$

we find that $\lambda\delta t < 2$, in order to decay to zero, ergo, the Euler method is said to be *conditionally* stable; the quality of the numerical solution depends on the step size $\delta t$.

## 7.4 The implicit Euler method

The Euler method we just discussed is called *explicit*, which means that we used a gradient at the current time step.

An implicit method uses a gradient at a future point. This seems a bit strange, because now we need to get a new value for $x$ with

$$x_{i+1} = x_i + f(x_{i+1}, t + \delta t)\delta t. \tag{7.11}$$

In the code we only need to make a small change:

```
function[x,t] = Euler(MyFunc,InitialValue,Start,Finish,Nsteps)
x(1) = InitialValue;
t(1) = Start;
dt = (Finish-Start)/Nsteps;
for i =1:Nsteps
 F = feval(MyFunc,x(i),t(i));
 t(i+1) = dt +t(i);
 x(i+1) = fsolve(@FunToSolve,x(i),[],x(i),t(i+1),MyFunc,dt);
end
t = t'; x = x';
return function residual = FunToSolve(x,xo,t,MyFunc,dt)
 residual = xo + feval(MyFunc,x,t)*dt-x;
return
```

If we now run the code with our test problem, for only 5 steps, we can see that the numerical solution is stable and more accurate than the explicit scheme (Figure 7.4).

## 7.5 Stability of the implicit Euler method

If we use an implicit method, we can make the same evaluation of stability as we did before:

$$\begin{aligned} x_{i+1} &= x_i + f(x_{i+1}, t + \delta t)\delta t && (7.12)\\ &= x_i + \lambda x_{i+1}\delta t, && (7.13) \end{aligned}$$

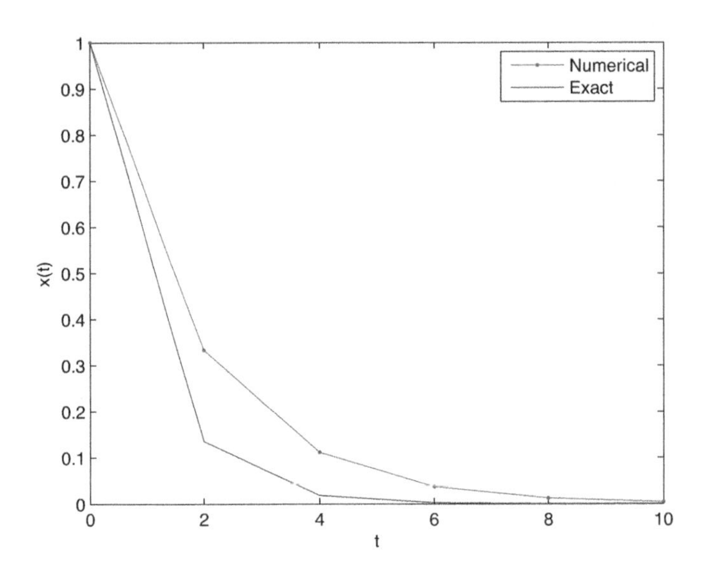


**FIGURE 7.4**
Only 5 steps with the implicit Euler scheme

which becomes after rewriting:

$$x_{i+1}(1 - \lambda \delta t) = x_i \rightarrow x_{i+1} = \frac{x_i}{(1 - \lambda \delta t)}. \tag{7.14}$$

The scheme becomes *unconditionally* stable when $\lambda < 0$ and conditionally stable if $\lambda > 0$.

## 7.6 Systems of ODEs

We often want to solve systems of coupled ODE's, like:

$$\frac{dx_1}{dt} = f_1(x_1, x_2, x_3, \cdots, t) \tag{7.15}$$

$$\frac{dx_2}{dt} = f_1(x_1, x_2, x_3, \cdots, t) \tag{7.16}$$

$$\frac{dx_3}{dt} = f_1(x_1, x_2, x_3, \cdots, t) \tag{7.17}$$

$$\cdots \tag{7.18}$$

We can easily adjust the program (the explicit version) for systems of ODE's.

```
function[x,t] = EulerCoupled(MyFunc,InitialValue,Start,Finish,
Nsteps)
x(:,1) = InitialValue';
t(1) = Start; dt = (Finish-Start)/Nsteps;
for i =1:Nsteps
 F = feval(MyFunc,x(:,i),t(i));
 t(i+1) = dt +t(i);
 x(:,i+1) = F*dt + x(:,i);
end
t = t'; x = x';
return
```

Similarly, you can adjust the implicit Euler code to deal with ODE systems. We can solve a linear ODE system, given as:

$$\frac{dx_1}{dt} = -x_1 - x_2 \tag{7.19}$$

$$\frac{dx_2}{dt} = x_1 - 2x_2, \tag{7.20}$$

with $x_1(0) = x_2(0) = 1$. First we write a function handle for this model as follows:

```
function [dxdt] = TestFunction2(x,t)
dxdt(1) = -1*x(1) - 1*x(2)
dxdt(2) = 1*x(1) - 2*x(2);
dxdt = dxdt'
return
```

And now you can solve the system by:

```
>>[x,t] = EulerCoupled(@TestFunction2,[1;1],0,10,100);
```

The results are plotted in the Figure 7.5. It should be noted that you need to supply a vector of initial values now.

## 7.7 Stability of ODE systems

Let us evaluate our example, written in matrix notation

$$\frac{d}{dt}\begin{bmatrix} x_1 \\ x_2 \end{bmatrix} = M \begin{bmatrix} x_1 \\ x_2 \end{bmatrix} \tag{7.21}$$

with:

$$M = \begin{bmatrix} -1 & -1 \\ 1 & -2 \end{bmatrix}. \tag{7.22}$$

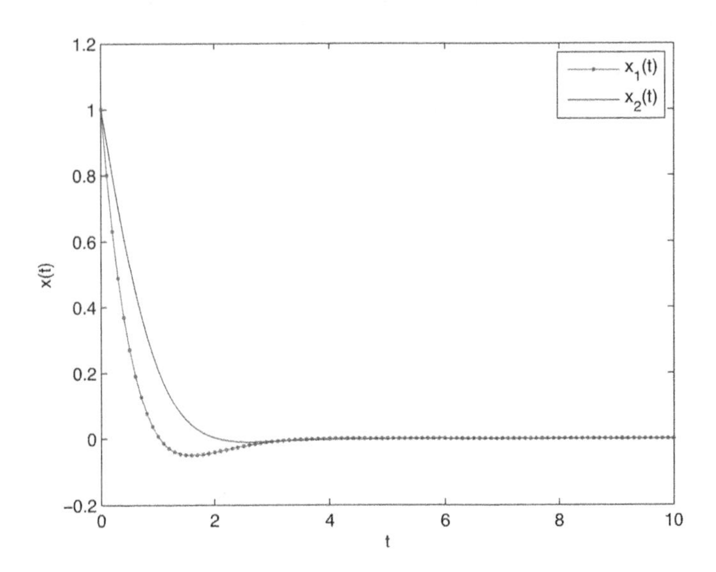


**FIGURE 7.5**
Numerical solution with Euler's method for the coupled ODE system

For linear ODE systems we can obtain an analytical solution by factorizing the $M$ matrix, as we discussed in Chapter 2:

$$M = U^{-1}\Lambda U. \tag{7.23}$$

We could rewrite the system as

$$\frac{d}{dt}\begin{bmatrix} x_1 \\ x_2 \end{bmatrix} = U^{-1}\Lambda U \begin{bmatrix} x_1 \\ x_2 \end{bmatrix}, \tag{7.24}$$

from which follows

$$U\frac{d}{dt}\begin{bmatrix} x_1 \\ x_2 \end{bmatrix} = UU^{-1}\Lambda U \begin{bmatrix} x_1 \\ x_2 \end{bmatrix}, \tag{7.25}$$

which equals:

$$\frac{d}{dt}\left(U\begin{bmatrix} x_1 \\ x_2 \end{bmatrix}\right) = \Lambda\left(U\begin{bmatrix} x_1 \\ x_2 \end{bmatrix}\right). \tag{7.26}$$

We can define a new set of variables $y$:

$$\begin{bmatrix} y_1 \\ y_2 \end{bmatrix} = U\begin{bmatrix} x_1 \\ x_2 \end{bmatrix} \tag{7.27}$$

and substitute Equation 7.27 into Equation 7.26:

$$\frac{d}{dt}\left(\begin{bmatrix} y_1 \\ y_2 \end{bmatrix}\right) = \Lambda\left(\begin{bmatrix} y_1 \\ y_2 \end{bmatrix}\right), \tag{7.28}$$

where $\Lambda$ was, of course, a diagonal matrix:

$$\frac{d}{dt}\begin{bmatrix} y_1 \\ y_2 \end{bmatrix} = \begin{bmatrix} \lambda_1 & 0 \\ 0 & \lambda_2 \end{bmatrix}\begin{bmatrix} y_1 \\ y_2 \end{bmatrix}. \tag{7.29}$$

This means the equations are no longer coupled, so we can find a solution:

$$\begin{bmatrix} y_1 \\ y_2 \end{bmatrix} = \begin{bmatrix} \alpha \exp(\lambda_1 t) \\ \beta \exp(\lambda_2 t) \end{bmatrix}, \tag{7.30}$$

from which follows, according to 7.27,

$$\begin{bmatrix} x_1 \\ x_2 \end{bmatrix} = U\begin{bmatrix} \alpha \exp(\lambda_1 t) \\ \beta \exp(\lambda_2 t) \end{bmatrix}, \tag{7.31}$$

where $U$ contains eigenvectors and the $\lambda$'s are eigenvalues. An eigenvalue can be a complex number. If the eigenvalues have an imaginary part, we can be sure that the system will oscillate. The real part of the eigenvalues determines whether a solution will go to a steady value or explode to infinity.

If we come back to our example, we will find for the $M$ matrix that the system has the following eigenvalues $\lambda_1 = -3/2+i\sqrt{3/2}$ and $\lambda_2 = -3/2-i\sqrt{3/2}$. The eigenvalues have an imaginary part, indicating that we may observe oscillation. Both real parts are negative, showing that the system will decay to a steady state.

We saw earlier that for the explicit Euler scheme $|1 + \lambda\delta t| < 1$ in order to be stable, but we can extend this to complex numbers, in which case the value for $|1 + \lambda\delta t|$ should be within a unit circle.

## 7.8 Stiffness of ODE systems

For nonlinear ODE systems, we have to look at the eigenvalues of the Jacobian in order to find out whether the system is stable. Here we also introduce stiffness as the ratio of the largest and smallest eigenvalue of the Jacobian. A system is called *stiff* if this ratio is much greater than unity. For stiff problems, you need an implicit solver!

## 7.9 Higher-order methods

Euler works reasonably well, but we could do better. There are more advanced methods that are based on the Euler method, but give better results. The main

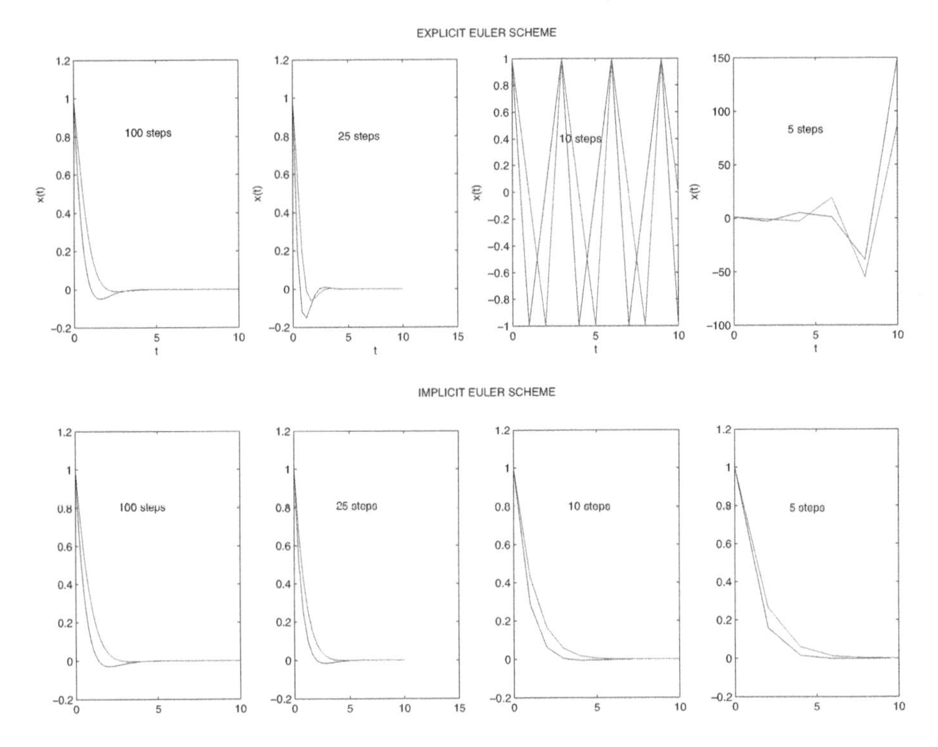


**FIGURE 7.6**
Some results for the explicit and implicit Euler's schemes for different numbers of steps

idea is that such methods use multiple points in their evaluation. One specific method that we are going to discuss shortly is the Runge-Kutta method. Figure 7.6 shows the numerical results for different integration steps.

A Runge-Kutta scheme uses weighted trajectories. First, we define a $k_i$:

$$k_i = f(x_i^{estimated}, t + c\delta t)\delta t, \tag{7.32}$$

where $0 < c < 1$. We can now update our solution by:

$$x_{\delta t} = x_0 + w_1 k_1 + w_2 k_2 + \cdots \tag{7.33}$$

The estimated values for $x$ at the $i$-th trajectory are then calculated from:

$$x_i^{estimated} = x_0 + a_{i,1} k_i \cdots \tag{7.34}$$

Thus, for example, for a second-order Runge-Kutta scheme, we need to evaluate two trajectories:

$$x_{\delta t} = x_0 + w_1 k_1 + w_2 k_2 + \cdots \tag{7.35}$$

where

$$k_1 = f(x_0, t)\delta t \tag{7.36}$$

and

$$k_2 = f(x_0 + a_{2,1}k_1, t + c_2\delta t)\delta t. \tag{7.37}$$

Now we must choose the weight ($w_1$,$w_2$, $a_{2,1}$) and the position $c_2$ such that we get an error of $O(\delta t^3)$ over a single time step.

If we expand $x$ over a Taylor series, we may find that

$$x = x_0 + f(w_1 + w_2) + w_w a_{2,1} f^x f \delta t^2 + w_2 c_2 f^i \delta t^2 + \emptyset(\delta t^3) \tag{7.38}$$

or

$$x = x_0 + f(x_0, t)\delta t + (f^x f(x_0, t) + f^i)\frac{\delta t^2}{2!} + \emptyset(\delta t^3). \tag{7.39}$$

We can use Equation 7.39 to obtain expressions for the weights. There are several options because we end up with 3 equations and 4 unknowns:

$$(w_1 + w_2) = 1 \tag{7.40}$$

$$w_2 a_{2,1} = 1/2 \tag{7.41}$$

$$w_2 c_2 = 1/2. \tag{7.42}$$

One option is, for example, the *Crank-Nicholson scheme*; by setting $c_2 = 1$, you can solve Equation 7.40 to $w_1 = w_2 = 1/2$ and $a_{2,1} = 1$.

Another option is called the *Euler mid-point scheme*, where with $c_2 = 1/2$ you will find $w_1 = 0$, $w_2 = 1$ and $a_{2,1} = 1/2$.

You could also derive a 4th-order RK scheme using the same approach. In MATLAB, the code for the RK4 scheme looks like this:

```
function [x,t] = RK4(MyFunc,InitialValues,Start,Finish,Nsteps)
x(:,1) = InitialValues'
 t(1) = Start; dt = (Finish - Start)/Nsteps;
for i = 1:Nsteps
 k1 = feval(MyFunc,x(:,i),t(i))*dt;
 k2 = feval(MyFunc,x(:,i) + k1/2,t(i) + dt/2)*dt;
 k3 = feval(MyFunc,x(:,i) + k2/2,t(i) + dt/2)*dt;
 k4 = feval(MyFunc,x(:,i)+k3,t(i) + dt)*dt;
 t(i+1) = dt + t(i);
 x(:,i+1) = (k1+2*k2+2*k3+k4)/6 + x(:,i);
end
t = t'; x = x';
return
```

In the 4th-order scheme you need to evaluate 4 trajectories. This type of integration scheme is accurate to $O(\delta t^4)$, so we would increase accuracy by a factor 10,000 if we reduce the time step with a factor 10. Figure 7.7 shows the comparison between Euler and Runge-Kutta. A Runge-Kutta scheme of 100 time steps would require a Euler scheme of 10,000,000 steps to get the same quality of the numerical solution.

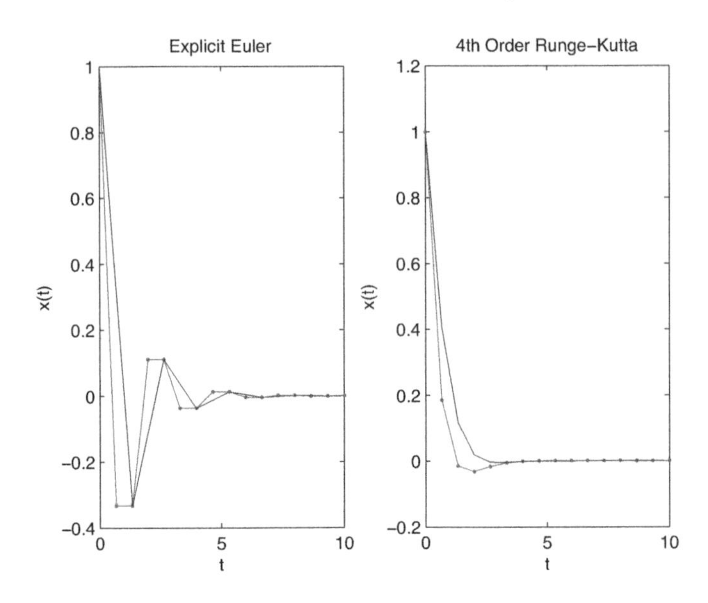


**FIGURE 7.7**
Euler versus Runge-Kutta for 15 time steps

## 7.10 Boundary value problems

We have only considered first-order differential equations until now. But there is a special class of differential equations that often occur when formulating engineering problems: the so-called boundary value problem (BVP), which is in fact a second-order differential equation.

Let us have a look at a second-order differential equation:

$$f\left(\frac{d^2y}{dx^2}, \frac{dy}{dx}, y, x\right) = 0 \tag{7.43}$$

Often, second-order differential equations can be rewritten into a system of first-order ordinary differential equations. Consider the second-order ODE:

$$\frac{d^2y}{dx^2} + \frac{q\,dy}{dx} = r \tag{7.44}$$

We can define a new variable $z$:

$$z = \frac{dy}{dx} \tag{7.45}$$

When substituting $z$ into the original equation, we get a system of two coupled ODEs:

$$\frac{dy}{dx} = z \tag{7.46}$$

$$\frac{dz}{dx} = r - qz \tag{7.47}$$

We learned that for initial value problems (with ODEs), the values of $y$ can be calculated by marching the solution forward in time from a starting point (the initial condition), with, for example, Euler's method. For boundary value problems (i.e., second-order differential equations), an initial condition is not sufficient. We have to define boundary conditions at more than one point.

To solve a boundary value problem, we can use the so-called shooting method. We first convert the BVP into a system of ODEs, then provide an initial guess for the unknown boundary condition. Subsequently, we solve the system and check if the resulting boundary condition is at its expected value. If not, we need to modify our guessed boundary value and solve again until we reach convergence. This iterative procedure is called shooting and can be done with a nonlinear equation solver.

As an example, consider a chemical reaction in a liquid film layer of thickness $\delta$, described by:

$$D\frac{d^2c}{dx^2} = kc \tag{7.48}$$

with $c(x = 0) = c_{aiL} = 1$ the interface concentration and $c(x = \delta) = 0$ the bulk concentration. Now we would like to solve the boundary value problem and plot the concentration profile in the film layer.

We first convert the BVP to a system of ordinary differential equations. We first define a flux $q$:

$$q = -D\frac{dc}{dx} \tag{7.49}$$

Now we find:

$$\frac{dc}{dx} = -\frac{1}{D}q \tag{7.50}$$

$$\frac{dq}{dx} = -kc \tag{7.51}$$

We know the boundary conditions for the concentration at $x = 0$ and $x = \delta$. The boundary condition for the flux at the interface is not known and should be solved iteratively.

The following MATLAB code will solve the problem using ODE45 and fzero:

```
function main

% Parameter definition
ps.D=1e-8;
ps.kR =10;
ps. delta =1e-4;

% Solve for flux boundary condition ( initial guess : 0)
opt = optimset ('Display ','iter ');
flux = fzero ( @RunBVP ,0,opt ,ps);

function f = RunBVP (bcq ,ps)
[x,cq] = ode45 (@BVPODE,[0 ps.delta ] ,[1 bcq], [], ps);
f = cq(end ,1) - 0;
plotyy (x,cq (: ,1) ,x,cq (: ,2));
xlabel('x [m]'); legend('c [ml/m^3]','q [ml/m^2/s]');
return ;


function dxdt = BVPODE (t,x,ps)
dxdt (1) = -1/ ps.D*x(2) ;
dxdt (2)= -ps.kR*x(1) ;
dxdt = dxdt';
return
```

The MATLAB program will generate the following output:

```
Search for an interval around 0 containing a sign change:
Func-count a          f(a)    b          f(b)     Procedure
   1       0          11.8334 0          11.8334  initial
                                                  interval
   3       -0.0282843 1066.45 0.0282843 -1042.79 search

Search for a zero in the interval [-0.0282843, 0.0282843]:
 Func-count     x             f(x)              Procedure
    3        0.0282843       -1042.79           initial
    4      0.000317363   2.12177e-09            interpolation
    5      0.000317363     3.1225e-16           interpolation
    6      0.000317363     3.1225e-16           interpolation
Zero found in the interval [-0.0282843, 0.0282843]
```

fzero has found a value of $q = 0.000317363$ for which $c(x = d) = 0$, as can be seen in Figure 7.8.

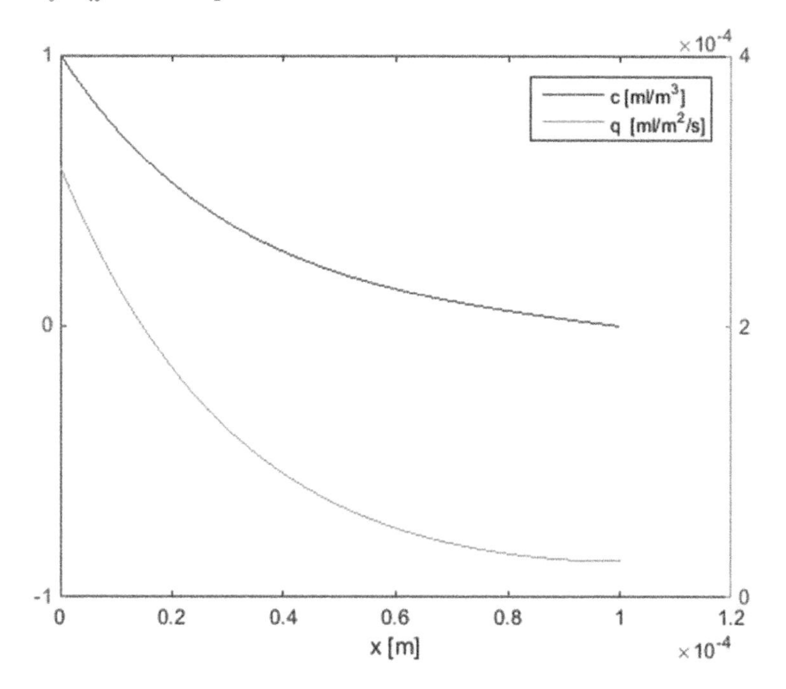


**FIGURE 7.8**
Concentration and flux profile over the interface

We can actually compute in this case an analytical value for $q$. It can be calculated from:

$$q = k_l E_A C_{Ail} \tag{7.52}$$

where the enhancement factor is given as:

$$E_A = \frac{Ha}{\tanh(Ha)} \tag{7.53}$$

and where the Hatta number is defined as:

$$Ha = \sqrt{\frac{kD}{k_l}} \tag{7.54}$$

The mass transfer coefficient can be determined from:

$$k_l = \frac{D}{\delta} \tag{7.55}$$

When filling out the data, we come to a similar number as our numerical value for $q$.

## 7.11 Summary

In this chapter you learned about the Euler method and what implicit and explicit schemes actually mean (current time, future time). We found out that methods can be (un)conditionally stable, and that we can use eigenvalues to tell us something about stability. Also, the term *stiffness* was introduced. Stiffness of a system tells you if you should use an explicit or implicit method to solve it (small or large time steps). We also looked at a higher-order method, the Runge-Kutta method. RK methods evaluate the solution over more than one trajectory point, and this means that approximations are more accurate, while requiring fewer time steps. We did not discuss it, but we will practice the MATLAB solvers for ODE's. Generally, ODE45 is used (on the basis of an RK scheme) and for stiff systems ODE15s is a good alternative.

## 7.12 Exercises

### Exercise 1

In the following ODE system:

$$\frac{dx_1}{dt} = -1x_1 - 1x_2 \tag{7.56}$$

$$\frac{dx_2}{dt} = 1x_1 - 2x_2, \tag{7.57}$$

with the initial conditions $x_1(0) = 1$ and $x_2(0) = 1$. Solve this system analytically, and factorize into U and $\Lambda$, calculating the eigenvector matrix and the eigenvalues of this system.

Is the solution stable? Explain.

Is the ODE system stiff or nonstiff? Explain.

### Exercise 2

We are going to write a MATLAB code to solve the ODE system of the previous equation with OD45. The first step is to write the ODE function:

```
function dxdt = odefun(t,x)
dxdt = zeros(2,1)
dxdt(1) = -1*x(1) - 1*x(2);
dxdt(2) = 1*x(1) - 2*x(2);
```

Now we can solve this system using ODE45 as follows:

```
options=odeset('RelTol',1e-4,'AbsTol',[1e-4 1e-4]);
Timespan = [0 10];
X0 = [0 1];
[T,X] = ode45(@odefun,Timespan,X0,options);
```

You can evaluate the solution graphically by:

```
plot(T,X(:,1),'-',T,X(:,2),'-.')
```

### Exercise 3

Let's evaluate a stiff problem:

$$\frac{dx_1}{dt} = -1x_1 - 1x_2 \tag{7.58}$$

$$\frac{dx_2}{dt} = 1x_1 - 5000x_2. \tag{7.59}$$

Use the routines from the previous exercise to solve the system. Add to the options:

```
options=odeset('RelTol',1e-4,'AbsTol',[1e-4 1e-4], 'Stats','on');
```

Write down the statistics. Replace the solver line with the following:

```
[T,X] = ode15(@odefun,Timespan,X0,options);
```

What are the differences between ODE45 and ODE15?

### Exercise 4

Integrate the following differential equations

$$\frac{dC_A}{dt} = -4C_A + C_B \tag{7.60}$$

$$\frac{dC_B}{dt} = 4C_A - 4_B \tag{7.61}$$

with $C_A(0) = 100$ and $C_B(0) = 0$. Integrate over the time period from $t =$ 1...5. Use Euler's method and a fourth-order Runge-Kutta method.

Which method would give a solution closer to the analytical solution? And why do the different methods give different results?

# 8

# Numerical integration

*In this chapter, we will learn how to calculate the value of an integral with three different numerical methods: Euler's method, the trapezoid method, and Simpson's rule. We will also learn what Richardson's correction is.*

## 8.1 Introduction

If we know of a function $f$ over a certain interval, the primitive $F$, we can easily determine the integral of that function over the interval by use of the main theorem of integration:

$$\int_a^b f(x)dx = F(b) - F(a) \tag{8.1}$$

However, sometimes it is not possible to determine the primitive of a function $f$. Alternatively, we can approximate the integral by numerical integration.

There are several numerical schemes for numerical integration, which we will discuss in this chapter.

## 8.2 Euler's method

Suppose we want to numerically determine the following integral:

$$I = \int_a^b f(x)dx \tag{8.2}$$

$a$ and $b$ are called the basic points, and the intersections with the vertical lines through $a$ and $b$ are called supporting points, as shown in Figure 8.1.

The surface of the area closed by the graph of $f$ and the $x$-axis and the vertical lines through the basic points is by definition equal to $I$. We might

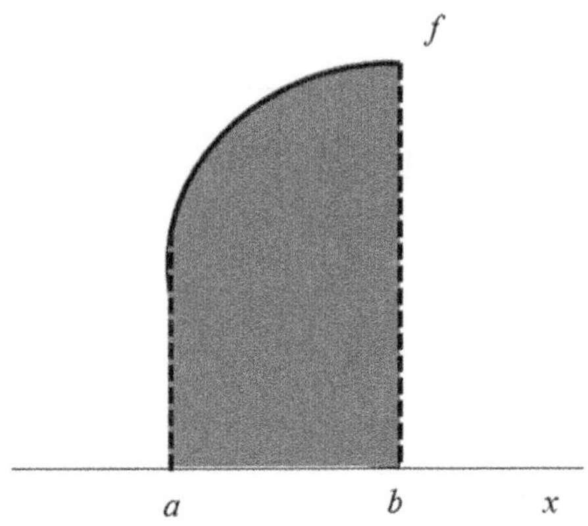


**FIGURE 8.1**
Surface below the graph of $f$ between $[a,b]$

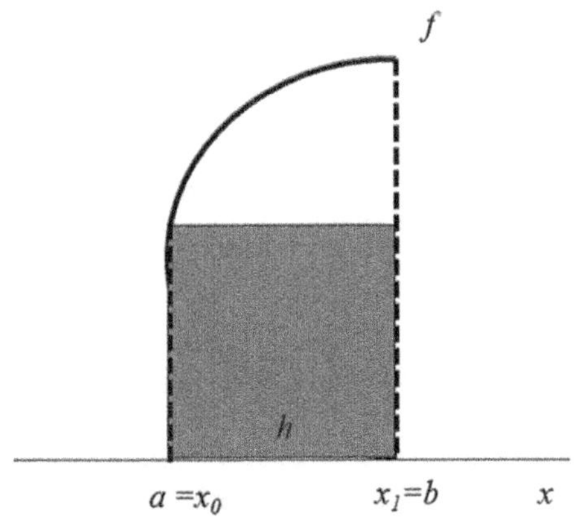


**FIGURE 8.2**
Approximation of the surface of Figure 8.1 through the surface of a rectangle

approximate $I$ by approximating the surface of a rectangle that appears by drawing a line perpendicular to the $x$-axis starting from the left support point, as shown in Figure 8.2.

It is obvious that for most cases, this approximation is very rough. In practice we can make the approximation much more accurate by dividing the interval $[a,b]$ into a number of connected subintervals and approximating each of the subintervals the same way. We should set the subintervals to equal lengths, such as $n$, which gives intervals with a length of $h = (b - a)/n$. If $x_0 = a$ and $x_n = b$, we can approximate $I$ like:

$$I \approx I_n^0 = h(f(x_0) + f(x_1) + \cdots + f(x_{n-1})) \tag{8.3}$$

This method is called Euler's method or the rectangle rule. If the integration interval is divided into more than one subinterval, we speak of the repeated rectangle rule. Figure 8.3 shows graphically how Euler's method works.

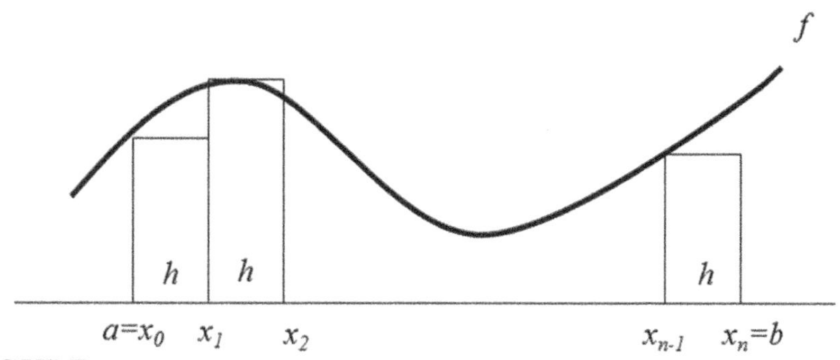


**FIGURE 8.3**
Application of Euler's method

Alternatively to Equation 8.3, we might have written:

$$I \approx I_n^0 = h(f(x_1) + f(x_2) + \cdots + f(x_n)) \tag{8.4}$$

In the case of Equation 8.4, the rectangles have heights that correspond to the function value at the right side of each of the subintervals. The approximation with Equation 8.4 will yield similar results as with Equation 8.3.

With the following code, the rectangle rule can be implemented easily in MATLAB:

```
function main
%% Small script integrating an equation with rectangle rule
for different sizes of subinterval h
for n = 1:100
    S(n) = rectanglerule(@func,4,6,n)
end

plot(1:n,S,'r.')
xlabel('number of integration steps, n');
ylabel('Numerical approximation of surface below function');

%% MATLAB function containing equation for integration
function f = func(x)
f = x.^3 + 2*x - 3;

%% MATLAB function containing the rectangle rule
function S = rectanglerule(func, a,b,n)

h = (b-a)/n;
x = a:h:b;
f = feval(func,x);
S= sum(h*f);
```

This MATLAB function solves $I = \int_4^6 (x^3 + 2x - 3)dx$ for different values of $n$. In Figure 8.4, the third-order polynomial is sketched, and in Figure 8.5, the numerical solution $S$ is plotted as a function of the number of integration steps $n$. For $n > 40$, the numerical solution no longer changes too much and convergence can be assumed.

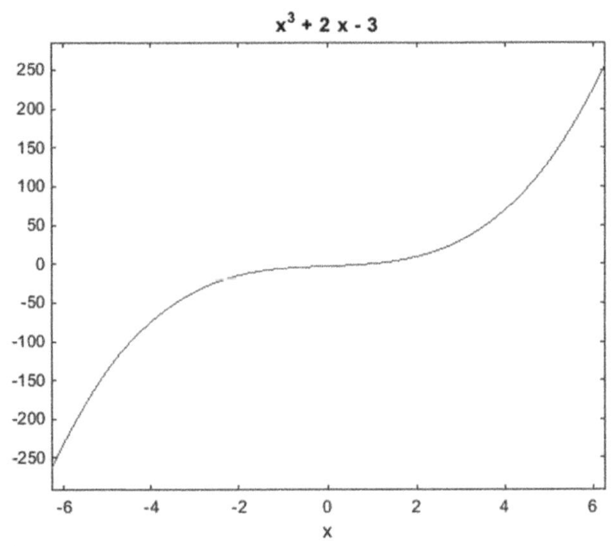


**FIGURE 8.4**
Plot of the function$f(x)$ for different values of $x$

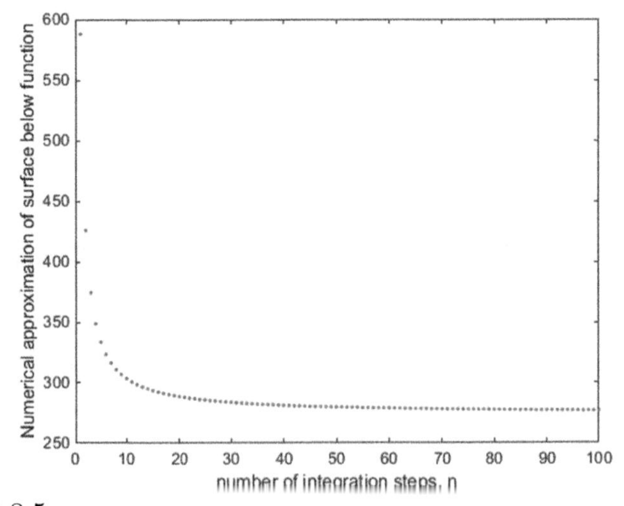


**FIGURE 8.5**
Plot of the numerical solution of the integral as a function of the number of integration steps

## 8.3 The trapezoid method

In the previous section, we saw that in the case of a monotonous continuous function, there can be two numerical approximations using the rectangle method. The method will always over- or underestimate the real solution. We might think of a better approximation, for example, by taking the average of the two versions of the rectangle method.

If we apply the two versions of the rectangle method to a function $f$ over one interval $[x_0, x_1]$ with $x_1 = x_0 + h$, and then take the geometric average of the two results, we will obtain:

$$I \approx I_1^1 = \frac{1}{2}(hf(x_0) + hf(x_1)) = \frac{h}{2}(f(x_0) + f(x_1)) \tag{8.5}$$

This result can be interpreted as the surface of a trapezoid that appears when we connect the support points with each other; see Figure 8.6.

To obtain a more accurate solution, the trapezoid rule can be repeated. Suppose we want to integrate the function of Figure 8.7 numerically with the trapezoid rule.

We divide the interval $[a,b]$ into $n$ subintervals, each with a length of $h = (b - a)/n$. We can now apply the trapezoid rule at each subinterval:

$$[a = x_0, x_1] \quad \frac{h}{2}(f(x_0) + f(x_1)) \tag{8.6}$$

$$[x_1, x_2] \quad \frac{h}{2}(f(x_1) + f(x_2)) \tag{8.7}$$

$$[x_{n-1}, x_n = b] \quad \frac{h}{2}(f(x_{n-1}) + f(x_n)) \tag{8.8}$$

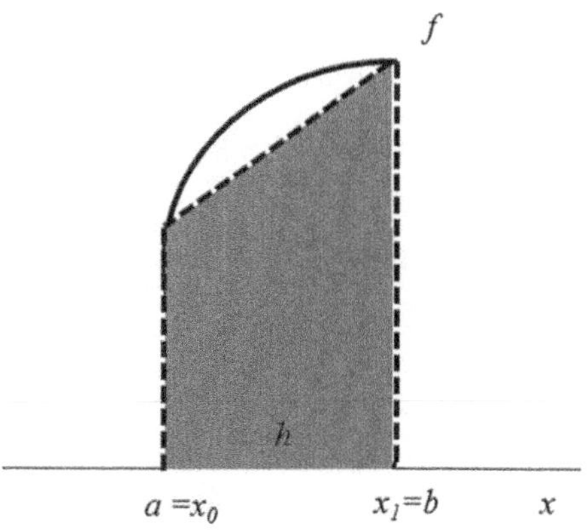


**FIGURE 8.6**
Surface below the graph of $f$ and above $[a,b]$

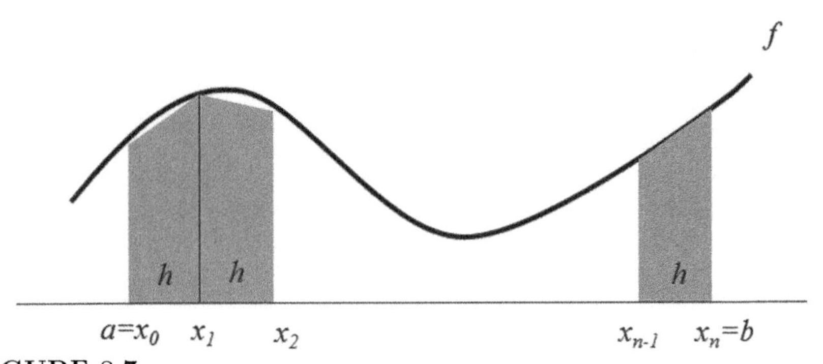


**FIGURE 8.7**
Repeated application of the trapezoid rule

If we add up all approximations, we will obtain:

$$I_n^1 = \frac{h}{2}(f(x_0) + 2f(x_1) + 2f(x_2) + \cdots + 2f(x_{n-1}) + f(x_n)) \tag{8.9}$$

Equation 8.9 is called the repeated trapezoid method.

## 8.4 Simpson's method

We will once more consider a function $f$ over the interval $[a,b]$. Instead of using the points $a$ and $b$, as we saw in the rectangle and trapezoid methods, we will now use three basic points: $a = x_0$, $(a+b)/2 = x_1$, and $x_2 = b$. In this case, $h = x_1 - x_0 = x_2 - x_1 = (x_2 - x_0)/2$.

One parabola fits exactly through the three support points; see Figure 8.8.

The parabola is given by the quadratic function:

$$p_2(x) = ax^2 + bx + c \tag{8.10}$$

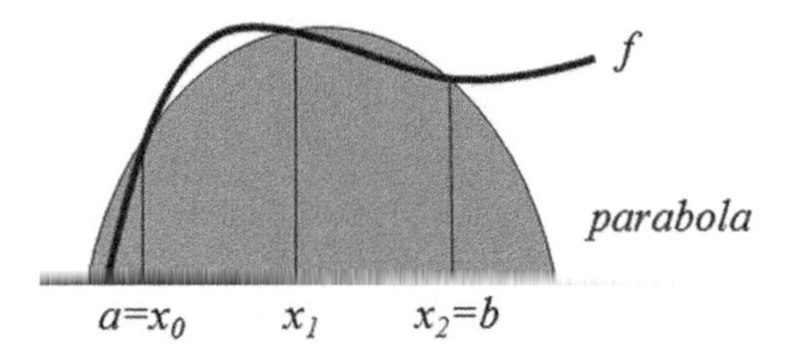


**FIGURE 8.8**
Approximation of the surface by calculating the surface below a parabola

As $f(x) = p_2(x)$, we can calculate:

$$\begin{aligned} f(x_0) &= ax_0^2 + bx_0 + c \\ f(x_1) &= ax_1^2 + bx_1 + c \\ f(x_2) &= ax_2^2 + bx_2 + c \end{aligned} \tag{8.11}$$

From this system of equations, $a$, $b$, and $c$ can be found in terms of $f$and $x$. After substitution, we find that:

$$I_2^2 = \int_{x_0}^{x_2=x_0+2h} p_2(x)dx = \frac{h}{3}(f(x_0) + 4f(x_1) + f(x_2)) \tag{8.12}$$

Equation 8.12 is called Simpson's rule. Just as we saw for the other two methods, Simpson's rule can be used in a repeated fashion, leading to higher accuracy.

Suppose that we want to use Simpson's rule in a repeated way. We divide the interval $[a,b]$ into $n/2$ equal subintervals $[a = x_0, x_2]$,$[x_2,x_4]$, $\ldots$, $[x_{n-2},x_n = b]$. It is noted that $n$ has to be an even number. On each of the subintervals, we can apply Simpson's rule, with $h = (x_n - x_0)/n$:

$$[x_0, x_2] \quad \frac{h}{3}(f(x_0) + 4f(x_1) + f(x_2)) \tag{8.13}$$

$$[x_2, x_4] \quad \frac{h}{3}(f(x_2) + 4f(x_3) + f(x_4)) \tag{8.14}$$

$$[x_{n-2}, x_n] \quad \frac{h}{3}(f(x_{n-2}) + 4f(x_{n-1}) + f(x_n)) \tag{8.15}$$

If we add all approximations, we obtain:

$$I_n^2 = \frac{h}{3}(f(x_0)+4f(x_1)+2f(x_2)+4f(x_3)+\cdots+2f(x_{n-2})+4f(x_{n-1})+f(x_n)) \tag{8.16}$$

Equation 8.16 is called the repeated Simpson's rule.

## 8.5 Estimation of errors using numerical integration

We understand that Euler's method (as well as the trapezoid and Simpson methods) only gives a numerical approximation to the real solution of the problem. In this section we will estimate how big this mismatch or error actually is. We will start with having a look at error development in Euler's method.

It can be proven that for a continuous function $f$, and with application of the rectangle rule on one interval, a number $\xi$ exists for which:

$$I = \int_{x_0}^{x_0+h} f(x)dx = hf(x_0) + \frac{h^2}{2}f'(\xi) \quad (x_0 < \xi < x_0 + h) \tag{8.17}$$

Here, $hf(x_0)$ is the approximation of the surface of a square and $(h^2/2)f'(\xi)$ is the so-called truncation error. From Equation 8.17 follows that the truncation error is proportional to $h^2$.

We have seen that the accuracy of Euler's method increases if the number of subintervals is increased. We are now interested in seeing how the error develops if we divide an interval $[a,b]$ into $n$ subintervals with equal lengths of $h = (b - a)/n$. By using Equation 8.17, we find:

$$\int_a^b f(x)\,dx = h(f(x_1) + f(x_2) + \cdots + f(x_n)) + (b - a)hf'(\xi) \quad (a < \xi < b) \tag{8.18}$$

The last term in Equation 8.18 is called the global truncation error for the repeated rectangle rule. The global error is proportional to $h$. That means that by reducing $h$ by a factor of 2, the error is halved. In practice, reducing $h$ with a factor of 2 also means that we need to calculate twice the number of function evaluations, which costs calculation time.

Worth noting is that the error in applying Simpson's rule for zero-, one-, or two-degree polynomials is zero. The numerical solution corresponds in this case to the exact solution.

## 8.6 The Richardson correction

In this section, we will show how integral approximations can be improved by using a correction that makes use of the result with only half the number of subintervals. This correction is called the Richardson correction. We will demonstrate it here for use with the rectangle method.

For a function $f$ that is differentiable over the interval $[a,b]$:

$$I = \int_a^b f(x)dx = I_n^0 + \frac{b-a}{2}\left(\frac{b-a}{n}\right)f'(\xi) \tag{8.19}$$

This follows directly from Equation 8.17. Here, $I_n^0$ is the numerical approximation of $I$ using the rectangle rule. When the first derivative of $f$ over $[a,b]$

is a constant function; that is, when $f'(x) = M_1$ holds true over the interval $[a,b]$, then the rest of the terms can be approximated in a smart way.

We can write Equation 8.19 for $n$ and $2n$ with $f'(\xi)$ approximated by $M_1$:

$$I \approx \int_a^b f(x)dx = I_n^0 - \frac{b-a}{2}\left(\frac{b-a}{n}\right)M_1 = I_n^0 + E_n^0 \tag{8.20}$$

$$I \approx \int_a^b f(x)dx = I_{2n}^0 - \frac{b-a}{2}\left(\frac{b-a}{n}\right)M_1 = I_{2n}^0 + E_{2n}^0 \tag{8.21}$$

If we subtract Equation 8.20 from Equation 8.21, we get:

$$0 \approx I_{2n}^0 - I_n^0 + E_{2n}^0 - E_n^0 \tag{8.22}$$

or:

$$I_{2n}^0 - I_n^0 \approx E_n^0 - E_{2n}^0 \tag{8.23}$$

By doubling the $n$, the error reduces by a factor of 2, and we can also write:

$$I_{2n}^0 - I_n^0 \approx 2E_{2n}^0 - E_{2n}^0 = E_{2n}^0 \tag{8.24}$$

Such that:

$$\int_a^b f(x)dx \approx I_{2n}^0 + \left(I_{2n}^0 - I_n^0\right) = 2I_{2n}^0 - I_n^0 \tag{8.25}$$

The correction term $I_{2n}^0 - I_n^0$ is called the Richardson correction.

As an example, suppose we will use the repeated rectangle rule to calculate the following integral:

$$\int_1^2 \frac{dx}{x}$$

For $n = 2$, 4, and 8, we obtain the following results:

$$\begin{aligned} I_2^0 &= 0.833333 \\ I_4^0 &= 0.759524 \\ I_8^0 &= 0.725372 \end{aligned}$$

The Richardson correction for $I_4^0 \rightarrow I_{2n}^0 - I_n^0 = -0.073809$.

Correcting now with the Richardson correction gives $I_4^0 = 0.759524 - 0.073809 = 0.68715$, which is closer to the exact solution of $I = 0.6931472$.

## 8.7 Summary

In this chapter, we discussed three methods for numerical integration: Euler's method, the trapezoid method, and Simpson's rule. We discussed Richardson's correction and showed its computational advantages, reducing calculation steps and increasing numerical accuracy.

## 8.8 Exercises

### Exercise 1

Calculate the value of the following integral with the repeated rectangle rule, while ensuring that the error is less then $0.5 \times 10^{-3}$:

$$\int_1^2 \frac{1}{1+x} dx$$

### Exercise 2

Calculate the value of the following integral with the repeated trapezoid rule for $n = 2$:

$$\int_0^1 (x^3 - x + 1) dx$$

Also calculate the exact outcome and compare with the numerical result.

### Exercise 3

The heat $Q$ generated when a current $i$ is flowing through a resistance $R$ during time $t$ can be calculated from:

$$Q = \int_{\tau=0}^{t} i^2 R \, d\tau$$

Use the following data to calculate the generated heat using the repeated Simpson's rule.

| $t$ (s) | 0 | 0.05 | 0.1 | 0.15 | 0.2 | 0.25 | 0.3 |
|---|---|---|---|---|---|---|---|
| $i^2 R$ (W) | 0.0 | 0.35 | 0.50 | 0.70 | 0.75 | 0.80 | |

### Exercise 4

The filling time $T$ of a vessel (in minutes) can be given by approximation of the following integral:

$$T = \int_{0.25}^{0.5} \frac{4500}{10 - 20h^{3/2}} dh$$

Compute the value of $T$.

# 9

# *Partial differential equations 1*

## 9.1 Introduction

In this chapter you will learn to compute numerical solutions to partial differential equations. There are several classes of differential equations: the parabolic, elliptic, and hyperbolic equations. We already saw the approach shown in this chapter in Chapter 5 on iterative methods. Now we are going to extend the method by transforming the PDE system into a system of ODE's.

## 9.2 Types of PDEs

The general structure of a second-order PDE can be given as

$$a\frac{\partial^2 u}{\partial t^2} = 2b\frac{\partial^2 u}{\partial t \partial x} + c\frac{\partial^2 u}{\partial x^2} = f(\frac{\partial u}{\partial t}, \frac{\partial u}{\partial x}, u). \tag{9.1}$$

We can define a determinant as

$$D = b^2 - 4ac. \tag{9.2}$$

In principle we can distinguish three classes of differential equations; if the determinant is smaller than zero, we call the equation *parabolic*, if it is equal to zero it is called *elliptic*, and if it is larger than zero it is called *hyperbolic*.

Often it is handy to identify whether a PDE has one- or two-way coordinates. Below are some examples of PDE's.

$$\frac{\partial T}{\partial t} + \frac{k}{\rho C_P}\frac{\partial^2 T}{\partial x^2} \tag{9.3}$$

Equation 9.3 is a parabolic equation. We see that time is a one-way coordinate and space is a two-way coordinate. That time is a one-way coordinate means that we can march a solution forward in time.

$$\frac{\partial^2 T}{\partial x^2} + \frac{\partial^2 T}{\partial x^2} = 0 \tag{9.4}$$

We got acquainted with the elliptic PDE of Equation 9.4 in the chapter on iterative methods. We saw that we have to solve for all values simultaneously.

We will not discuss hyperbolic equations, for example, the wave equation:

$$\frac{\partial^2 u}{\partial t^2} - c\frac{\partial^2 u}{\partial x^2} = 0. \tag{9.5}$$

If we want to solve a PDE, we need to convert it to a system of algebraic equations, or ODEs. The first thing we need to do is discretize the spatial domain. There are several ways to discretize space, by means of *finite differences*, *finite volumes*, and *finite elements*.

## 9.3 The method of lines

In this chapter we will have a look at the method of lines. We will look at the unsteady-state Laplace equation for heat conduction in a slab.

In the method of lines, we discretize that spatial domain of the PDE to produce a set of ODE's, which govern the temperature at each point in the solution. We will use finite differences to accomplish this.

The unsteady-state heat conduction equation is given by

$$\frac{\partial T}{\partial t} = \alpha \nabla^2 T \tag{9.6}$$

In two dimensions this equation can be written as

$$\frac{\partial T}{\partial t} = \alpha \left( \frac{\partial^2 T}{\partial x^2} + \frac{\partial^2 T}{\partial y^2} \right). \tag{9.7}$$

We want to solve this equation over the domain shown in Figure 9.1 The first thing we do is to place a grid over the solution domain and track temperature at each POINT of the grid. We will use an equal grid space. The index $k$ can be obtained from $i$ and $j$:

$$k = i + Nx(j-1), \tag{9.8}$$

such that each temperature at grid point $(i, j)$ can be subsequently represented by $T_{i,j}$:

$$T_{i,j} = T_{k=i+Nx(j-1)}. \tag{9.9}$$

We can discretize the heat equation to find an ODE that governs the temperature at node $k$, so we discretize:

$$\frac{\partial T_{i,j}}{\partial t} = \alpha \left( \left.\frac{\partial^2}{\partial x^2}\right|_{i,j} + \left.\frac{\partial^2 T}{\partial y^2}\right|_{i,j} \right). \tag{9.10}$$

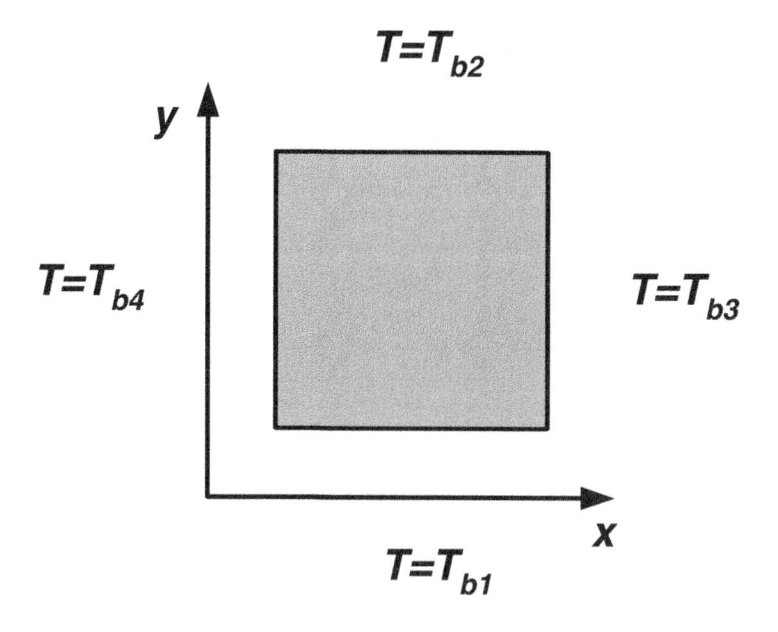


**FIGURE 9.1**
Spatial domain of the steel slab

As you saw in Chapter 5 on iterative methods, we can obtain a discretization using finite differences by assuming a piecewise linear relation:

$$\frac{\partial^2 T}{\partial x^2} \approx \frac{\left.\frac{\partial T}{\partial x}\right|_{i+1/2} - \left.\frac{\partial T}{\partial x}\right|_{i-1/2}}{\Delta x} \tag{9.11}$$

$$\approx \frac{\frac{T_{i+1,j}-T_{i,j}}{\Delta x} - \frac{T_{i,j}-T_{i-1,j}}{\Delta x}}{\Delta x} \tag{9.12}$$

$$= \frac{T_{i+1,j} - 2T_{i,j} + T_{i=1,j}}{\Delta x^2}. \tag{9.13}$$

We do the same thing for the y-direction and we find after rewriting and replacing the $i$ and $j$ with $k$:

$$\frac{dT_k}{dt} = \alpha \left( \frac{T_{k+1} - 2T_k + T_{k-1}}{\Delta x^2} + \frac{T_{k+Nx} - 2T_k + T_{k-Nx}}{\Delta y^2} \right). \tag{9.14}$$

The boundaries have a fixed temperature; they do not obey our discretized equations. We can incorporate the boundaries by eliminating the temperatures at the boundary nodes from the set of equations, or we could write $dT_{boundary}/dt = 0$, or we could include boundary node equations as a set of algebraic equations.

We are going to use the first option, and we will eliminate the temperatures at the boundary nodes from the equations. We know that for a node on a

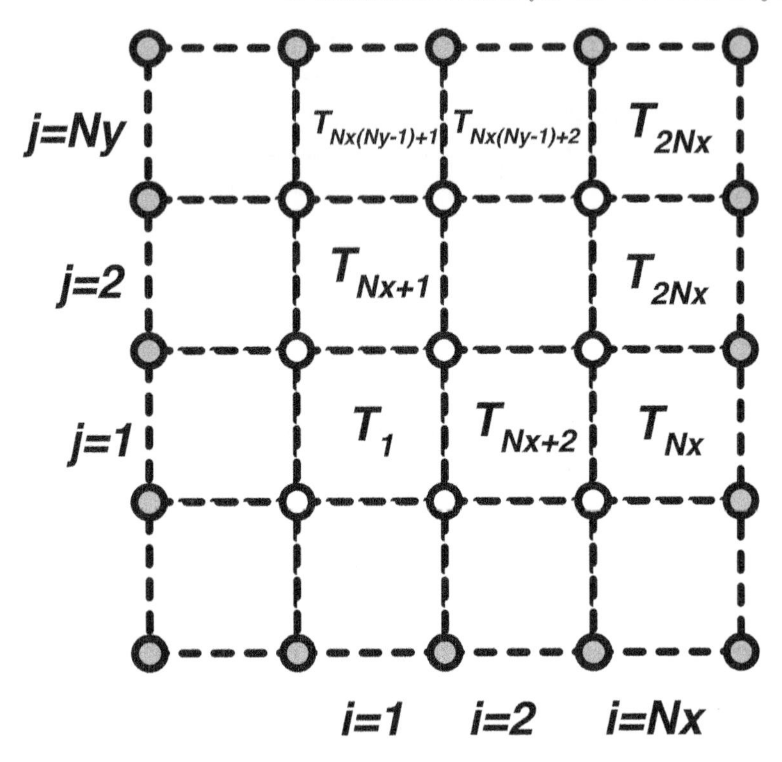


**FIGURE 9.2**
The grid space with elimination of the boundary nodes from the equation

border we can write (when $\Delta x = \Delta y$):

$$\frac{dT_k}{dt} = \frac{\alpha}{\Delta x^2}\left(T_{k+1} - 2T_k + T_{k+Nx} - 2T_k + T_{k-Nx}\right). \tag{9.15}$$

Figure 9.2 shows the proposed grid spacing for the steel slab. You could check the grid space and set the nodes on the borders as written above. This is how we eliminate the boundary conditions; the node on border 1: $T_{k-Nx} = T_{b1}$, on border 2: $T_{k+Nx} = T_{b2}$, on border 3: $T_{k-1} = T_{b3}$ and on border 4: $T_{k+1} = T_{b4}$.

After elimination of the boundaries, we end up with the complete model as given in

$$\frac{dT}{dt} = AT + b, \tag{9.16}$$

or

$$\frac{d}{dt}\begin{bmatrix} T_1 \\ T_2 \\ T_3 \\ T_4 \\ T_5 \\ T_6 \\ T_7 \\ T_8 \\ T_9 \\ T_{10} \\ T_{11} \\ T_{12} \end{bmatrix} = \frac{\alpha}{\Delta x^2}\begin{bmatrix} -4 & 1 & 0 & 1 & 0 & 0 & 0 & 0 & 0 & 0 & 0 & 0 \\ 1 & -4 & 1 & 0 & 1 & 0 & 0 & 0 & 0 & 0 & 0 & 0 \\ 0 & 1 & -4 & 0 & 0 & 1 & 0 & 0 & 0 & 0 & 0 & 0 \\ 1 & 0 & 0 & -4 & 1 & 0 & 1 & 0 & 0 & 0 & 0 & 0 \\ 0 & 1 & 0 & 1 & -4 & 1 & 0 & 1 & 0 & 0 & 0 & 0 \\ 0 & 0 & 1 & 0 & 1 & -4 & 0 & 0 & 1 & 0 & 0 & 0 \\ 0 & 0 & 0 & 1 & 0 & 0 & -4 & 1 & 0 & 1 & 0 & 0 \\ 0 & 0 & 0 & 0 & 1 & 0 & 1 & -4 & 1 & 0 & 1 & 0 \\ 0 & 0 & 0 & 1 & 0 & 1 & 0 & 1 & -4 & 0 & 0 & 1 \\ 0 & 0 & 0 & 1 & 0 & 0 & 1 & 0 & 0 & -4 & 1 & 0 \\ 0 & 0 & 0 & 1 & 0 & 0 & 0 & 1 & 0 & 1 & -4 & 1 \\ 0 & 0 & 0 & 1 & 0 & 0 & 0 & 0 & 1 & 0 & 1 & -4 \\ 0 & & & & & & & & & & & \end{bmatrix}$$

$$\times \begin{bmatrix} T_1 \\ T_2 \\ T_3 \\ T_4 \\ T_5 \\ T_6 \\ T_7 \\ T_8 \\ T_9 \\ T_{10} \\ T_{11} \\ T_{12} \end{bmatrix} + \frac{\alpha}{\Delta x^2}\begin{bmatrix} T_{b1} + T_{b3} \\ T_{b1} \\ T_{b1} + T_{b4} \\ T_{b3} \\ 0 \\ T_{b4} \\ T_{b3} \\ 0 \\ T_{b4} \\ T_{b2} + T_{b3} \\ T_{b2} \\ T_{b2} + T_{b4} \end{bmatrix} \tag{9.17}$$

We can now solve this system of ODE's with, for example, the ODE45 solver of MATLAB, which is straightforward. In Figure 9.3 the outcome is shown. You see profiles for $x$ and $y$ at different times. We may define a time constant

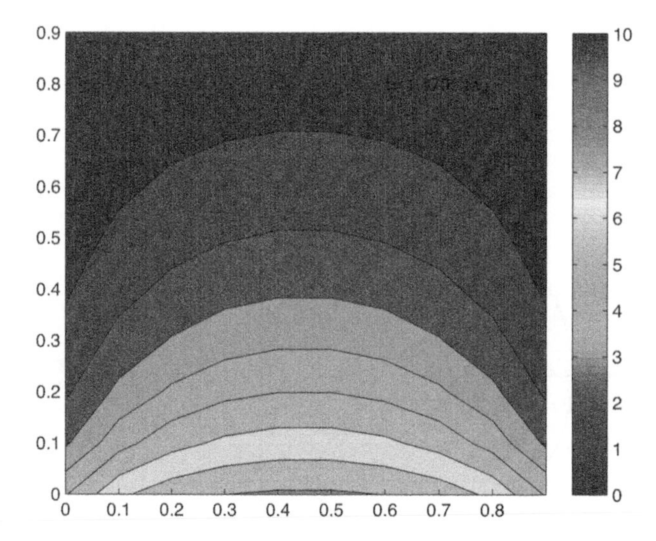


**FIGURE 9.3**
The time constant $\tau = L^2/\alpha$ gives you an idea how long the simulation should be run; $L$ is the characteristic length, defined as $L = (Nx + 1)\Delta x/2$

as the ratio of the characteristic length of the slab squared and the thermal diffusivity. Using half this length will give you a good idea of how long you should run the simulation.

## 9.4 Stability

You could check the stability of a system by evaluating the Jacobian. The eigenvalue of the Jacobian with the largest magnitude determines which time step you can take. For our system, the Jacobian equals the $A$ matrix.

As we discussed before, we can use Gershgorin's theorem to estimate eigenvalues as

$$|\lambda - m_{k,k}| \leq |m_{k,1}| + |m_{k,2}| + \cdots + |m_{k,N-2}| + |m_{k,N-1}| + |m_{k,N}| . \quad (9.18)$$

We find that all eigenvalues have to be within the circle for the solution to be stable. If we zoom in a little in the Argand plot, as shown in Figure 9.4, we actually find that the eigenvalues should be within a radius of $4\alpha/\Delta x^2$, from the point $(-4\alpha/\Delta x^2, 0)$. The worst possible case would be to have eigenvalues of $-4\alpha/\Delta x^2$. If you interrupt the simulation half way and calculate the eigenvalues of A (`>>eig(A)`), you will come close to this value.

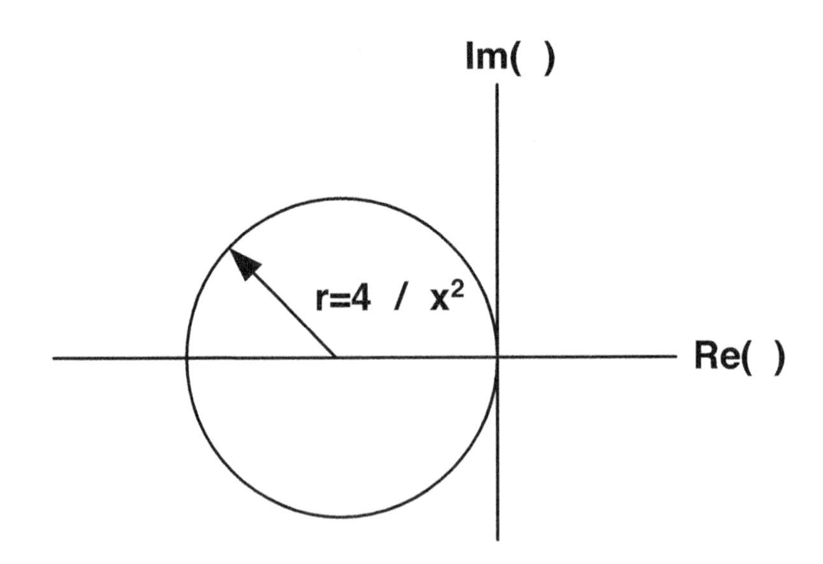


**FIGURE 9.4**
The Argand diagram; solving PDEs with explicit schemes means that we have a maximum step size we can take in order to obtain a stable solution

If we had used the explicit Euler method to step forward in time, we would have found that the method is only stable when

$$-\lambda\delta t < 2 \rightarrow \delta t < 2\Delta x^2/8\alpha. \tag{9.19}$$

This outcome highlights a problem we encounter in solving a PDE: If we use an explicit time stepping scheme, we will find a maximum step size we can take in order to create a stable solution.

## 9.5 Summary

In this chapter we saw that we can convert PDE's to a system of ODE's. Although not explicitly mentioned this time, we see that we are confronted with sparse $A$ and $J$ matrices. We can use the eigenvalues of the Jacobian to investigate stability of the method, and we see that when we are using implicit methods to solve the ODE system we are limited by a maximum step size we can take to keep the solution stable.

## 9.6 Exercises

### Exercise 1

Open the function `HeatConduction.m` and see if you can make it work for $T_{b1} = 5$, $T_{b2} = 10$, $T_{b3} = 10$, $T_{b4} = 10$, $T_0 = 0$ and $\alpha = 1$.

### Exercise 2

MATLAB has a solver for initial-boundary value problems for parabolic-elliptic PDEs in one dimension. In this exercise we are going to get acquainted with the specific solver `pdepe`.

We are going to compute and plot the solution of the following PDE:

$$\pi^2 \frac{\partial u}{\partial t} = \frac{\partial}{\partial x}\left(\frac{\partial u}{\partial x}\right). \tag{9.20}$$

This equation holds onto the interval $0 \leq x \leq 1$ for times $0 \leq t$. The PDE satisfies the initial condition $u(x,0) = sin\pi x$ and the boundary conditions $u(0,t) = 0$ and $\pi e^{-t} + \frac{\partial u}{\partial x}(1,t) = 0$.

The PDE solver has a convention for the formulation of the PDE, the initial conditions, and the boundaries. Generally, the PDE should be formulated in terms of $c, f, s$, and $m$:

$$c\left(x,t,u,\frac{\partial u}{\partial x}\right)\frac{\partial u}{\partial t}=x^{-m}\frac{\partial}{\partial x}\left(x^m f\left(x,t,u,\frac{\partial u}{\partial x}\right)\right)+s\left(x,t,u,\frac{\partial u}{\partial x}\right). \quad (9.21)$$

where $a \leq x \leq b$, $t_0 \leq t \leq t_f$, and $m = 0, 1, 2$. The initial conditions should be written as $u(x, t_0) = u_0(x)$ and for the boundary condition the following form is required:

$$p(x,t,u)+q(x,t)f\left(x,t,u,\frac{\partial u}{\partial x}\right)=0. \quad (9.22)$$

Now we have to write functions for the PDE `pdex1pde`, the initial condition `pdex1ic`, and the boundaries `pdex1bc`:

```
function [c,f,s] = pdex1pde(x,t,u,DuDx)
c = ...;
f = ...;
s = ...;
```

Fill in the dots, and formulate the functions for the initial conditions and the boundaries yourself.

What is the value for $m$?

We will solve the PDE using the following code:

```
sol = pdepe(m,@pdex1pde,@pdex1ic,@pde1bc,x,t);
```

where `x=linspace(0,1,20)` and `t=linspace(0,2,5)`

Make a surface plot of the solution and give a profile of $u$ as a function of $x$ at $t = t_f$.

### Exercise 3

Express the two-dimensional parabolic partial differential equation

$$\frac{\partial u}{\partial t}=\alpha\left(\frac{\partial^2 u}{\partial x^2}+\frac{\partial^2 u}{\partial y^2}\right) \quad (9.23)$$

in an explicit finite difference formulation. Determine the limits of conditional stability for this method.

# 10

# *Partial differential equations 2*

## 10.1 Introduction

In this chapter we are going to look at finite volumes to solve transport PDEs. After the formulation of a general transport PDE and introduction of finite volumes, we will try to solve such equations and we will find out that convection and diffusion play a significant role in the solution of the PDE.

## 10.2 Transport PDEs

For any property $\phi$ per mass unit, the following transport equation holds:

$$\frac{\partial \rho\phi}{\partial t} + \nabla \cdot J = b, \tag{10.1}$$

where $\rho$ is the density and $b$ is the production rate of $\phi$. $J$ is the total flux and can be expressed as:

$$J = \rho u \phi - D\nabla\phi, \tag{10.2}$$

where $u$ is the velocity and $D$ the diffusivity. We can combine Equations 10.1 and 10.2 to get

$$\frac{\partial \rho\phi}{\partial t} + \nabla \cdot (\rho u \phi - D\nabla\phi) = b. \tag{10.3}$$

Equation 10.3 is parabolic; we can derive it from a balance: ACCUMULATION = IN − OUT + PRODUCTION. If we evaluate a control volume, with a volume $dV = dxdydx$, as shown in Figure 10.1, our balance would look like this:

$$\frac{\partial \rho\phi}{\partial t} dV = \left(J_x - J_{x+dx}\right) dzdy + \left(J_y - J_{y+dy}\right) dxdz + \left(J_z - J_{z+dz}\right) dxdy + bdV. \tag{10.4}$$

If we take the difference in flux as infinitesimally small, we can rewrite as:

$$\frac{\partial \rho\phi}{\partial t} dV = -\frac{\partial J_x}{\partial x} dxdydz - \frac{\partial J_y}{\partial y} dxdydz - \frac{\partial J_z}{\partial z} dxdydz + bdV. \tag{10.5}$$

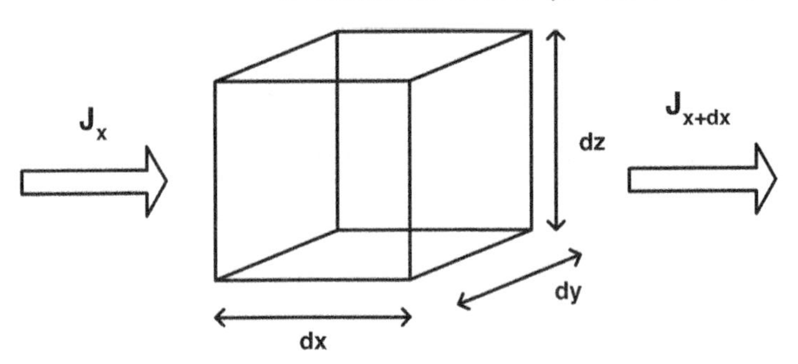


**FIGURE 10.1**
Control volume for our balance

Now we can divide by $dV$ and obtain

$$\frac{\partial \rho\phi}{\partial t} = -\frac{\partial J_x}{\partial x} - \frac{\partial J_y}{\partial y} - \frac{\partial J_z}{\partial z} + b = \nabla \cdot J + b. \tag{10.6}$$

## 10.3 Finite volumes

We could consider a volume, not necessarily a cube, and integrate the transport equation over this volume $V$, which would give us

$$\int_V \left( \frac{\partial \rho\phi}{\partial t} + \nabla \cdot J \right) dV = \int_\nu b dV. \tag{10.7}$$

For a fixed volume, the following holds:

$$\frac{\partial}{\partial t} \int_V \rho\phi dV + \int_V \nabla \cdot J dV = \int_\nu b dV. \tag{10.8}$$

We can use Gauss's theorem to convert this volume integral into a surface integral

$$\int_V \nabla \cdot J dV = \int_S n \cdot J dS, \tag{10.9}$$

where $n$ is the unit normal vector at each point. If we combine Equations 10.8 and 10.9, we get the following result:

$$\frac{\partial}{\partial t} \int_V \rho\phi dV = \int_\nu b dV - \int_S n \cdot J dS, \tag{10.10}$$

in which the left-hand side is the rate of $\phi$ accumulating in the control volume and the right-hand side is the production rate minus the rate at which $\phi$ flows across the boundaries. In other words, material leaving the volume will flow into another, i.e., material is not lost, ergo, $\phi$ is conserved.

## 10.4 Discretizing the control volumes

We can discretize the control volumes and index the cells with $i, j$, and $k$. Figure 10.2 shows how volumes can be discretized in space. If we now write out a discretized balance for control cell $i, j, k$ we would obtain:

$$\begin{aligned}\frac{\partial \rho \phi_{(i,j,k)}}{\partial t} &= J_{x(i-1,j,k)} \Delta y \Delta z + J_{y(i,j-1,k)} \Delta x \Delta z \\ &\cdots + J_{(z(i,j,k-1)} \Delta x \Delta y - J_{x(i,j,k)} \Delta y \Delta z \\ &\cdots - J_{y(i,j,k)} \Delta x \Delta z - J_{z(i,j,k)} \Delta x \Delta y \\ &\cdots + b \Delta x \Delta y \Delta z.\end{aligned}$$

We now need an expression for the fluxes across the faces of the control volume:

$$J_{x(i,j,k)} = \rho u \phi_{i+1/2} - D \left. \frac{\partial \phi}{\partial x} \right|_{i+1/2} . \quad (10.11)$$

The first part of the RHS is the convective term, and the second part is the diffusive term.

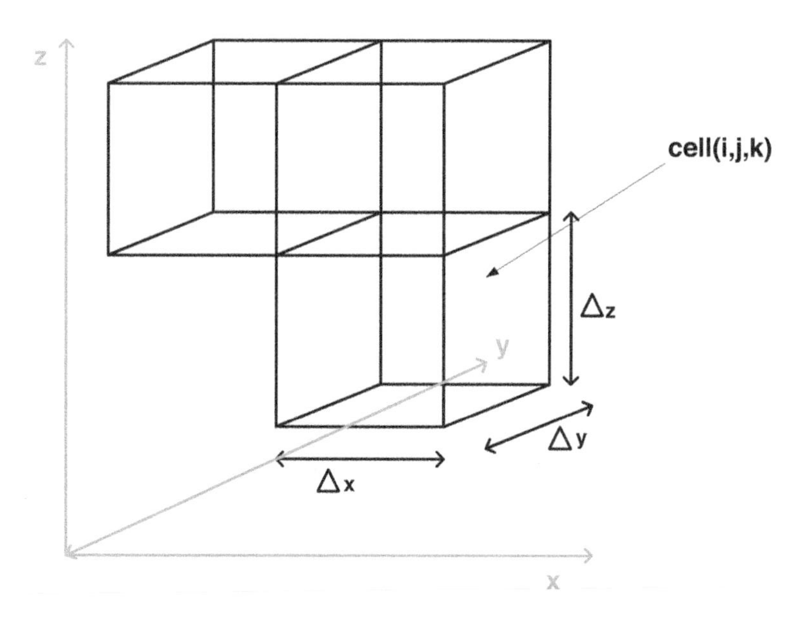


**FIGURE 10.2**
Discretization of the control volumes

Actually, we would like to have the difference in flux between the $i + 1/2$ and $i - 1/2$ interface

$$\left(J_{x(i-1,j,k)} - J_{x(i,j,k)}\right) = (\rho u_x)_{i-1/2}\phi_{i-1/2} - (\rho u_x)_{i+1/2}\phi_{i+1/2} - D\left.\frac{\partial \phi}{\partial x}\right|_{i-1/2} + D\left.\frac{\partial \phi}{\partial x}\right|_{i+1/2}. \tag{10.12}$$

We can use linear approximations for the diffusive and convective terms of the transport equation, using so-called *central differences*. For the diffusive terms,

$$\left.\frac{\partial \phi}{\partial x}\right|_{i+1/2} = \frac{\phi_{i+1} - \phi_i}{\Delta x}$$

$$\left.\frac{\partial \phi}{\partial x}\right|_{i-1/2} = \frac{\phi_i - \phi_{i-1}}{\Delta x},$$

from which follows:

$$-D\left.\frac{\partial \phi}{\partial x}\right|_{i-1/2} + D\left.\frac{\partial \phi}{\partial x}\right|_{i+1/2} = D\frac{\phi_{i+1} - 2\phi_i + \phi_{i-1}}{\Delta x}. \tag{10.13}$$

For the convective terms we can write

$$(\rho u_x)_{i-1/2}\phi_{i-1/2} - (\rho u_x)_{i+1/2}\phi_{i+1/2}, \tag{10.14}$$

or, when using central differences,

$$(\rho u_x)_{i-1/2}\frac{\phi_i - \phi_{i-1}}{2} - (\rho u_x)_{i+1/2}\frac{\phi_{i+1} - \phi_i}{2}. \tag{10.15}$$

Now we have formulated the overall system in discretized equations. Let us have a look at an example.

## 10.5 Transfer of heat to fluid in a pipe

Let us consider heat transfer to a fluid flowing in a tube with a small diameter, as shown in Figure 10.3. For the first $L_h$ meters, the tube is heated by raising the outside temperature to $T_w$, which gives a flux $q$ into the fluid ($h$ is the heat transfer coefficient). The last part of the tube is insulated. The PDE that describes this system is given as

$$\rho C_p \frac{\partial T}{\partial t} = -u\rho C_p \frac{\partial T}{\partial z} + k\frac{\partial^2 T}{\partial z^2} + h(T_w - T)\frac{p}{A} \tag{10.16}$$

with the following boundary conditions:

$$\left.\frac{\partial T}{\partial t}\right|_{z=L} = 0 \tag{10.17}$$

$$T|_{z=0} = T_0. \tag{10.18}$$

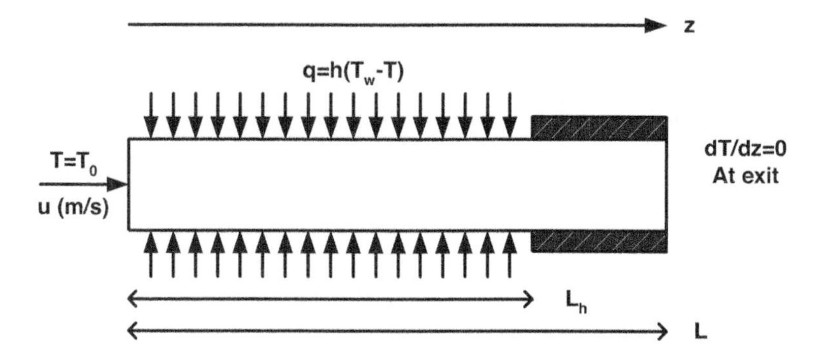


**FIGURE 10.3**
Heat conduction in a tube

We can divide the domain into small volumes as shown in Figure 10.4. For a control volume $i$, the flux across the interface is given as:

$$J_i = \rho u C_p T_{i+1/2} - k \left.\frac{\partial T}{\partial t}\right|_{z_{i+1/2}}. \tag{10.19}$$

Now we can draw the heat balance for the control volume as:

$$A\delta z \rho C_p \frac{dT_i}{dt} = AJ_{i-1} - AJ_i + \delta z hp(T_w - T) \tag{10.20}$$

with

$$J_i = \rho u C_p T_{i+1/2} - k \left.\frac{\partial T}{\partial t}\right|_{z_{i+1/2}},$$

which can be reformulated for the diffusive term

$$\left.\frac{dT}{dz}\right|_{z_{i+1/2}} = \frac{T_{i+1} - T_i}{\delta z} \tag{10.21}$$

and the use of central differences in the convective term

$$T_{i+1/2} = \frac{T_{i+1} + T_i}{2}. \tag{10.22}$$

Combining Equations 10.20, 10.21, and 10.22 leads to

$$\rho C_p \frac{dT_i}{dt} = \rho u C_p \frac{T_{i-1} + T_{i+1}}{2\delta z} + k\frac{T_{i+1} - 2T_i + T_{i-1}}{\delta z^2} + \frac{hp}{A}(T_w - T_i). \tag{10.23}$$

You should note that for the control volumes that are insulated $h = 0$. Of course, the boundary conditions should also be discretized. At the inlet the discretized boundary condition is

$$T_1 = T_0 \rightarrow \frac{dT_1}{dt} = 0 \tag{10.24}$$

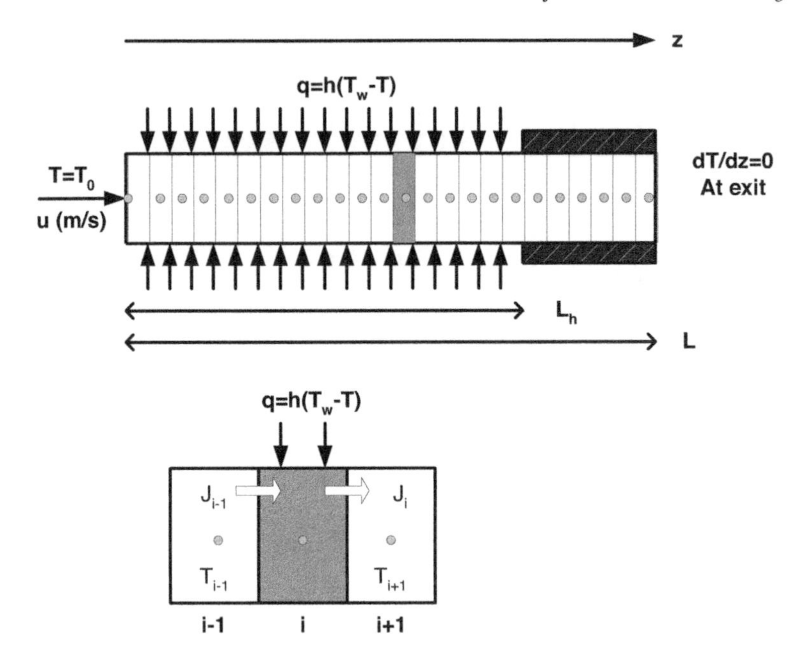


**FIGURE 10.4**
Tube divided into finite volumes

and at the outlet we will have the following discretization:

$$A\delta z\rho C_p\frac{dT_N}{dt} = A\rho uC_pT_N + A\rho uC_p\frac{T_N+T_{N-1}}{2} + kA\frac{T_N-T_{N-1}}{\delta z} + \delta zhp(T_w - T_N). \tag{10.25}$$

This boundary condition is a so-called *upwind approximation*. The third term of the right-hand side differs from our original Equation 10.23, as Equation 10.23 always depends on a discretization at the $i+1$ cell. With the upwind approximation we enforce an end to our equation system.

We can now write all discretized equations as the following ODE system:

$$\frac{dT}{dt} = MT + B, \tag{10.26}$$

where $M$ is a sparse banded matrix:

$$M = \begin{bmatrix} 0 & 0 & 0 & 0 & \cdots & 0 \\ b & a & c & 0 & \cdots & 0 \\ 0 & b & a & c & \cdots & 0 \\ 0 & 0 & b & a & c & 0 \\ \vdots & \vdots & \vdots & \vdots & \ddots & 0 \\ 0 & 0 & 0 & 0 & b & d \end{bmatrix} \tag{10.27}$$

in which

$$a = \frac{1}{\rho C_p}\left(-2\frac{2k}{\delta z^2} - \frac{hp}{A}\right) \quad (10.28)$$

$$b = \frac{1}{\rho C_p}\left(\frac{k}{\delta z^2} + \frac{\rho u C_p}{2\delta z}\right), \quad (10.29)$$

and where $c = d = b$.

$$B = \frac{phT_w}{A\rho C_p}(0, 1, \cdots, 1)^T \quad (10.30)$$

It could be convenient to render our parabolic PDE as dimensionless. We could, for example, use a dimensionless, time, distance, and time coordinate, respectively:

$$\vartheta = \frac{T - T_w}{T_0 - T_w} \quad (10.31)$$

$$\eta = \frac{z}{L} \quad (10.32)$$

$$t' = \frac{t}{\tau}, \quad (10.33)$$

where $\tau$ is the time constant given as

$$\tau = \frac{\rho C_p L^2}{k}. \quad (10.34)$$

We now can rewrite our transport equation as

$$\frac{\partial \vartheta}{\partial t'} = Pe\frac{\partial \vartheta}{\partial \eta} - \frac{\partial^2 \vartheta}{\eta^2} = -Bi\vartheta, \quad (10.35)$$

where $Pe$ is the *Peclet* number, which defines the rate of convection versus the rate of conduction as

$$Pe = \frac{uL\rho C_p}{k}, \quad (10.36)$$

and $Bi$ is the *Biot* number, which gives the ratio of internal heat transfer and external heat transfer according to

$$Bi = \frac{L^2 hp}{kA}. \quad (10.37)$$

## 10.6 Simulation of the heat PDE

If we solve the system using MATLAB you may obtain a result as shown in Figures 10.5 and 10.6, in which temperature is given as a function of the

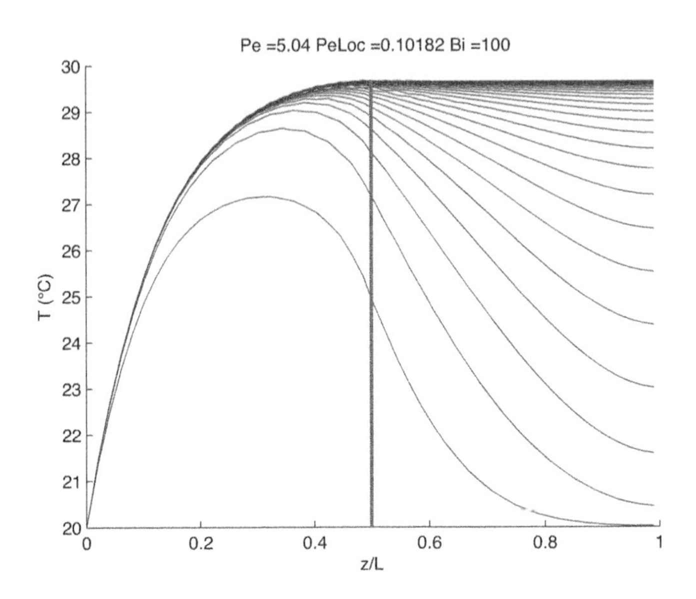


**FIGURE 10.5**
Numerical solution to the PDE where $k = 0.01$

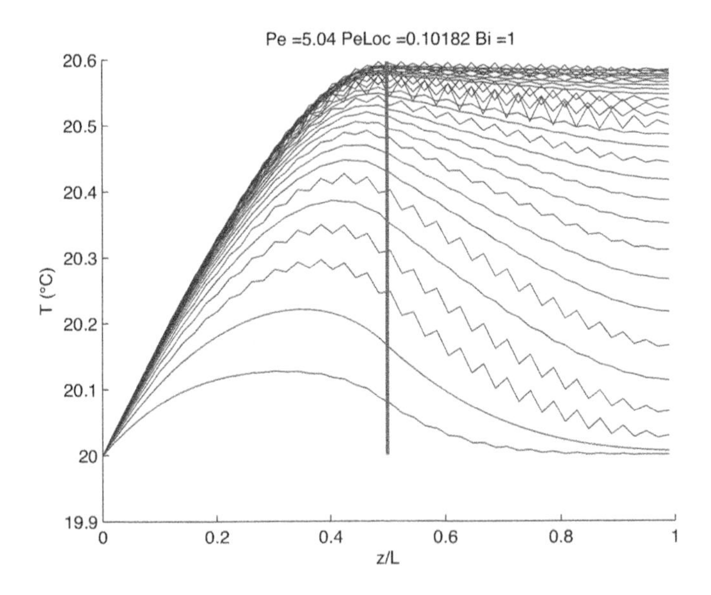


**FIGURE 10.6**
Numerical solution to the PDE where $k = 1.00$

distance for different times. The Peclet number for both cases differs. As the Peclet number increases, difficulties with the simulations arise, i.e., for processes where convection is a dominant factor, the system is increasingly difficult to solve accurately.

## 10.7 Summary

All transport equations have the same parabolic form, which means that the solution can be marched forward in time. With the finite volume method we conserve $\phi$. We evaluated the finite volume method with a practical case (heat transfer in a tube) and saw that the Peclet number influences the result. For higher Peclet numbers, the system is increasingly difficult to solve. If the process is governed by convection, you may expect problems and the only way to deal with that is by decreasing the size of the control volumes and time steps.

## 10.8 Exercises

### Exercise 1

Open the function `Tube_PDE.m` and see if you can run it. Try it for different settings of the grid and spacing and play a little with the physical parameters.

What do you observe, especially in relation to the Peclet number?

### Exercise 2

We are going to practice a bit more with the function `pdepe` to test what you remember from previous chapters. It's going to be a bit more difficult as we're now going to solve a system of PDEs:

$$\begin{aligned} \frac{\partial u_1}{\partial t} &= 0.024\frac{\partial^2 u_1}{\partial x^2} - F(u_1 - u_2) \\ \frac{\partial u_2}{\partial t} &= 0.17\frac{\partial^2 u_2}{\partial x^2} + F(u_1 - u_2), \end{aligned}$$

where $F(y) = \exp(5.73y) - \exp(-11.46y)$ and $y = u_1 - u_2$.

Initial conditions are

$$\begin{aligned} u_1(x,0) &= 1 \\ u_2(x,0) &= 0, \end{aligned}$$

and the boundary conditions are:

$$\begin{aligned} \frac{\partial u_1}{\partial x}(0,t) &= 0 \\ u_2(0,t) &= 0 \\ u_1(1,t) &= 1 \\ \frac{\partial u_2}{\partial x}(1,t) &= 0. \end{aligned}$$

Write functions for the PDE system `pdex2pde.m`, the initial conditions `pdex2ic.m`, and the boundaries `pdex2bc.m`.

Solve the system with `pdepe` over the interval $x = 0 \cdots 1$ and $t = 0 \cdots 2$.

Evaluate the profiles for $u_1$ and $u_2$ in a surface plot.

# 11

# *Data regression and curve fitting*

## 11.1 Introduction

In this chapter we are going to learn how to fit measurement data to a model using the least squares method. We will also discuss something about error and accuracy in data fitting.

## 11.2 The least squares method

Figure 11.1 shows a plot of some experimental data, where we measured $y$ as a function of $x$. We would like to fit a third-order polynomial of the following form to this data:

$$\hat{y} = a_1 + a_2 x + a_3 x^2 + a_4 x^3. \tag{11.1}$$

We could write the model as a product of a matrix and a vector:

$$\hat{y} = Xa. \tag{11.2}$$

The $X$ matrix is often called the *design matrix* and the vector $a$ contains the fit parameters. Equation 11.2 is actually given as

$$\begin{bmatrix} \hat{y}_1 \\ \hat{y}_2 \\ \hat{y}_3 \\ \vdots \\ \hat{y}_N \end{bmatrix} = \begin{bmatrix} 1 & x_1 & x_1^2 & x_1^3 \\ 1 & x_2 & x_2^2 & x_2^3 \\ 1 & x_3 & x_3^2 & x_3^3 \\ \vdots & \vdots & \vdots & \vdots \\ 1 & x_N & x_N^2 & x_N^3 \end{bmatrix} \begin{bmatrix} a_1 \\ a_2 \\ a_3 \\ a_4 \end{bmatrix}. \tag{11.3}$$

$N$ is the total number of points. We now define an absolute error, sometimes called the *residual* as

$$d_i = (y_i - \hat{y}_i), \tag{11.4}$$

which is the difference of the measured data and the predicted model.

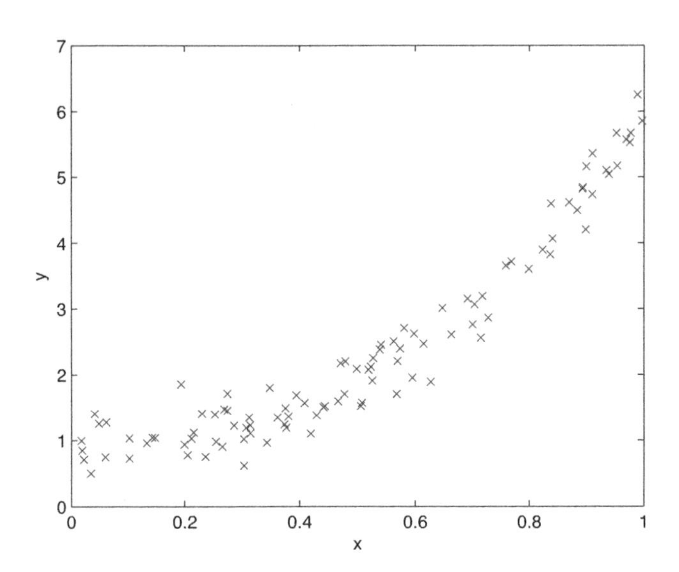


**FIGURE 11.1**
Experimental data

We could now define the sum of squared errors as

$$\sum_i (d_i)^2 = \sum_i (y_i - \hat{y}_i)^2 , \tag{11.5}$$

or in vector notation:

$$\sum_i (d_i)^2 = d \cdot d = d^T \times d = (y_i - \hat{y}_i)^T (y_i - \hat{y}_i) . \tag{11.6}$$

Now we need to determine values for the fit parameters that minimize the sum of the squared errors. That means that we will take the partial derivative with respect to each fit parameter and set these gradients to zero, according to

$$\frac{\partial}{\partial a_j} \left[ (y^T - (Xa)^T)(y - Xa) \right] = 0, \tag{11.7}$$

which can be rewritten as

$$\frac{\partial}{\partial a_j} \left[ (y^T - X^T a^T))(y - Xa) \right] = 0. \tag{11.8}$$

After application of the product rule we obtain

$$(y^T - X^T a^T) X \frac{\partial}{\partial a_j} [(a)] + \frac{\partial}{\partial a_j} \left[ (a)^T \right] X^T (y - Xa) = 0. \tag{11.9}$$

The partial derivatives with respect to $a$ are actually the unit vector

$$(y^T - X^T a^T)Xe_j + e_j^T X^T(y - Xa) = 0, \tag{11.10}$$

which can be rewritten as

$$(y - Xa)^T Xe_j + e_j^T X^T(y - Xa) = 0, \tag{11.11}$$

following

$$(Xe_j)^T(y - Xa) + e_j^T X^T(y - Xa) = 0. \tag{11.12}$$

After rearrangement we get

$$2X^T(y - Xa) = 0, \tag{11.13}$$

from which follows

$$X^T y = X^T Xa, \tag{11.14}$$

from which we can take an expression for $a$:

$$a = (X^T X)^{-1} X^T y. \tag{11.15}$$

Equation 11.15 is the outcome of the linear *least squares method.* It shows you how to obtain the model parameters on the basis of your $x$ and $y$ data.

If we have the same number of data points as fit parameters, we have a linear system

$$\begin{bmatrix} \hat{y}_1 \\ \hat{y}_2 \\ \hat{y}_3 \\ \hat{y}_4 \end{bmatrix} = \begin{bmatrix} 1 & x_1 & x_1^2 & x_1^3 \\ 1 & x_2 & x_2^2 & x_2^3 \\ 1 & x_3 & x_3^2 & x_3^3 \\ 1 & x_4 & x_4^2 & x_4^3 \end{bmatrix} \begin{bmatrix} a_1 \\ a_2 \\ a_3 \\ a_4 \end{bmatrix}, \tag{11.16}$$

which could be solved in MATLAB with `a=X\y`. If there are more data points ($N > 4$), we can write an analogue, but maybe a consistent solution does not exist (the system is overspecified). However, the backslash operator in MATLAB will always find values for the vector $a$ that minimize the sum of squares.

For our data shown in Figure 11.1, we could type this at the command prompt:

```
>> N = length(x)
>>X(:,1) = ones(N,1)
>>X(:,2) = x;
>>X(:,3) = x.^2
>>X(:,4) = x.^3
>>a = X\y
```

Figure 11.2 shows the regressed third-order polynomial and the data.

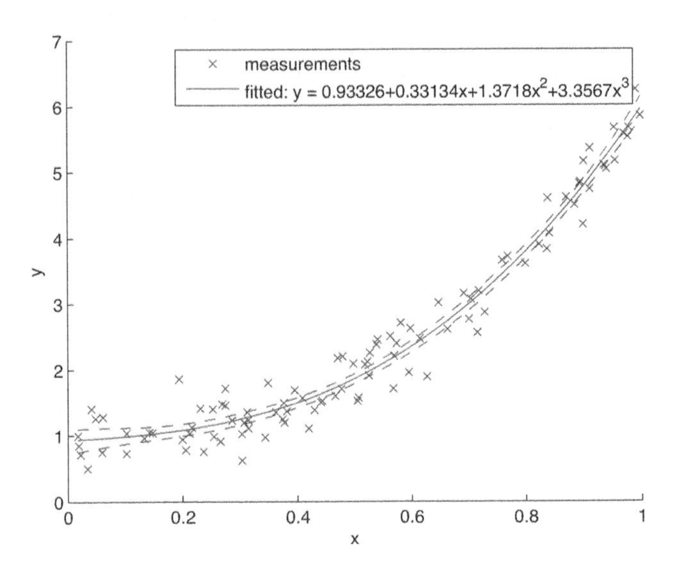


**FIGURE 11.2**
Experimental data and regressed polynomial

## 11.3 Residual analysis

After fitting the unknown parameters, we must ask how good the model actually is. We defined the absolute error, $d$, earlier. If the same experiment had been repeated many times, a random error would occur, which means a (normal) distribution of $y$ values would be produced together with a distribution of errors.

For a model to make sense, the data points should be scattered randomly around the model predictions: the mean of the error should be zero. You should check how the error evolves and if there is or is not a correlation with the measured value. If the former is the case, it probably indicates that something is wrong with your model structure. Figure 11.3 shows the residual plot for our example. You could have an indication of how well your model performs on the basis of variance determination. We recognize three types of variances: variance in the data, in the residuals, and in the model itself. For the data (sometimes called regression sum of squares or SSR):

$$\sigma_y^2 = \frac{1}{N}\sum(y_i - \bar{y})^2, \qquad (11.17)$$

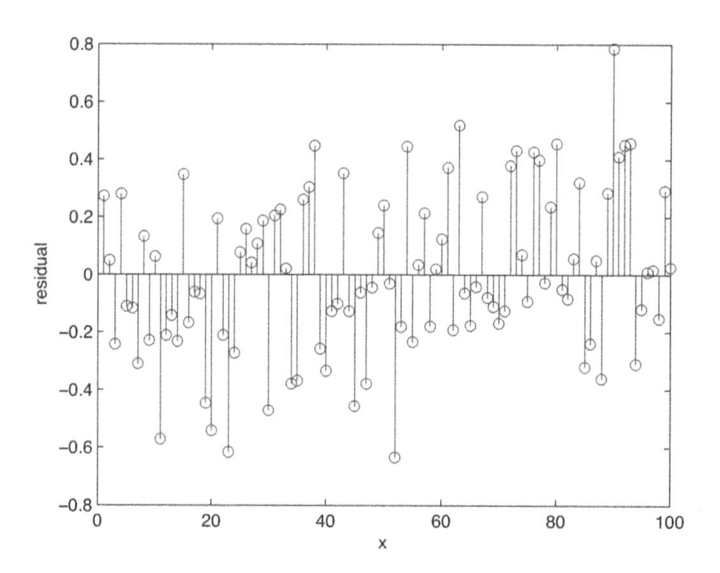


**FIGURE 11.3**
Residual plot

for the residual (sometimes called sum of squared errors or SSE):

$$\sigma_{error}^2 = \frac{1}{N}\sum (d_i)^2, \tag{11.18}$$

and for the model (sometimes called sum of squares total or SST):

$$\sigma_{model}^2 = \frac{1}{N}\sum (\hat{y}_i - \bar{\hat{y}})^2. \tag{11.19}$$

Given that the error is uncorrelated (random), we can state that

$$\sigma_y^2 = \sigma_{error}^2 + \sigma_{model}^2. \tag{11.20}$$

The correlation coefficient $R^2$ is the ratio of the variance of the model and the variance of the data:

$$R^2 = \frac{\sigma_{model}^2}{\sigma_y^2} = 1 - \frac{\sigma_{error}^2}{\sigma_y^2}. \tag{11.21}$$

The closer its absolute value comes to unity, the better your model is.

An uncorrelated error (mean will be zero) means that SSE, SST, and SSR will have $\chi^2$-distributions and the ratios will have an F-distribution. If SSR/SSE is large, the model is good. There is a chance that the model is rubbish, but that SSR/SSE will yield a good value. Analysis of variance (ANOVA) will be a good tool to calculate the probability of such a thing happening.

## 11.4 ANOVA analysis

The ANOVA will calculate F-values on the basis of the degrees of freedom and the sum of squared errors to see whether certain factors are significant or not. Large values for F indicate that the probability that your prediction is good on the basis of a bad model is very small. In fact, this ANOVA may be identified as hypothesis testing, where we accept or reject the null hypothesis of there being no correlation between $x$ and $y$.

The following ANOVA table is obtained for our example:

| Source | Deg. Freedom | Sum of squares | F-value |
|---|---|---|---|
| Regression | $K = 4$ | SSR = 224.44 | F= (SSR/4)/(SSE/95) |
| Residual | $N - K - 1 = 95$ | SSE = 8.103 | = 657.84 |
| Total | $N - 1 = 99$ | SST = 232.55 | |

## 11.5 Confidence limits

The confidence limits for the model parameters give us an indication of the reliability of the estimated model parameters. We can calculate the confidence limits for the model parameters from the t-distribution, according to

$$a_j - t\frac{\sigma^2_{error}}{\nu}\left[(X^TX)^{-1}\right]_{j,j} \le a_j \le a_j + t\frac{\sigma^2_{error}}{\nu}\left[(X^TX)^{-1}\right]_{j,j}, \tag{11.22}$$

where $\nu$ are the degrees of freedom. The values for $t$ need to be looked up from tables. For example, for a 95% confidence interval with 96 degrees of freedom, the value for $t = 1.98$.

Similarly, we can calculate confidence limits for each of the points that we have predicted:

$$\hat{y}_i - t\frac{\sigma^2_{error}}{\nu}\left[\sqrt{X(X^TX)^{-1}X^T}\right]_{j,j} \le \hat{y}_i \le \hat{y}_j + t\frac{\sigma^2_{error}}{\nu}\left[\sqrt{X(X^TX)^{-1}X^T}\right]_{j,j}. \tag{11.23}$$

See also Constantinides and Mostoufi (1999) for a fuller discussion.

## 11.6 Summary

In this chapter we have seen how fit parameters of a model can be fitted to a data set using the linear least squares method. We found out how to calculate the regression coefficients and how to perform a statistical analysis of the model using ANOVA. We also postulated expressions for the confidence limits for the fit parameters and the predicted points.

## 11.7 Exercises

### Exercise 1

In this exercise we are going to use the `fit` routine to fit data to the following function:

$$y(x) = \frac{ax}{ax + b}. \tag{11.24}$$

We want to determine the unknown fit parameters $a$ and $b$. Data was stored in the file `xydataset1.mat`. In MATLAB you can load data simply by typing `load xydataset1`; similarly, you can save data by typing `save xydataset1`.

After you load the data set, you have to define what kind of fit options you want to use and how your fit model should look:

```
s = fitoptions('Method','NonlinearLeastSquares',...
'Lower',[0,0],...
'Upper',[Inf,Inf],...
'Startpoint',[1 1]);.
```

You can define the fit model with:

```
f = fittype('a*x/(a*x+b)','options',s);
```

To actually calculate the fit parameters, the only command you need to type is

```
[model,stats] = fit(x,y,f)
```

Plot the data and evaluate the statistical data supplied. What is SSE, RMSE, etc.?

### Exercise 2

In this exercise we wish to determine the fit parameters $k_1$ and $k_2$ of the following simple ODE model:

$$\frac{du}{dt} = -k_1 u + k_2, \tag{11.25}$$

with the initial condition $u(0) = 1$.

To find the values for $k_1$ and $k_2$, we are going to use `lsqnonlin`. The first thing you have to do, of course, is to load the data `load tudataset1`.

`lsqnonlin` normally has the following structure:

```
k = lsqnonlin(fun,k0,lb,ub,options)
```

Thus, you need to supply initial guesses for your fit parameters ($k_0$), lower and upper search bounds (lb and ub), and an option set, normally something like:

```
options = optimset('TolX',1.0E-6,'MaxFunEvals',1000);
```

where you specify the tolerance and the maximum number of function evaluations.

`lsqnonlin` also requires you to define a minimization criterion (fun), something like,

$$f = (\hat{y} - y)^2, \tag{11.26}$$

so there is a quadratic difference between model and experimental data.

Another nasty thing here is that you are not dealing with an algebraic equation, but with a differential equation. Somehow you need to incorporate a function that describes the ODE function and solves it with an ODE solver.

Thus, the question is how to determine the fit parameters and the confidence limits? Show your fitted model and the data in a plot.

### Exercise 3

See if you can find a way to solve the isotherm of Equation 10.24 with two alternative MATLAB tools: `cftool` and `nlinfit`.

# 12

# *Optimization*

## 12.1 Introduction

In this chapter we are going to look at several techniques to solve optimization problems. Optimization problems occur anywhere in chemical engineering, all to improve operation or design of process systems.

Generally an optimization problem has the following structure:

$$
\begin{aligned}
\min f(x) & \\
s.t. & \\
& g(x) = 0 \\
& h(x) \geq 0
\end{aligned}
$$

in which $f(x)$ is the goal function, or objective function, that should be minimized (it should be noted that $\min f(x) = \max -f(x)$) and where $g(x)$ form the equality constraints and $h(x)$ the inequality constraints that should be satisfied. $f$, $g$, and $h$ can be linear or nonlinear and the variable $x$ could be discrete or continuous. Depending on the structure, there are several techniques available to solve the resulting problem.

Two properties of the functions $f, g, h$ are very important and will tell you if the problem can be solved easily or not. First, there is *continuity*. Optimization models that have functions with discontinuities are harder to solve. Often, optimization is concerned with studying the derivatives of the functions and the derivative at a discontinuity does not exist. Also, discrete functions are discontinuous, e.g., pipe diameters that can be employed in the construction of a plant.

Convexity of the functions that you are evaluating will also tell you how easily an optimization problem can be solved. A function is called *convex* if at any two points on the function, a straight line can be drawn that does not cross the function. Non-convex problems are hard to solve.

In the following sections, we will discuss several methods for solving optimization problems.

## 12.2 Linear programming

In linear programming (LP), the objective function and the constraints are linear functions, for example:

$$\max z = f(x_1, x_2) = 40x_1 + 88x_2$$
$$s.t.$$
$$2x_1 + 8x_2 \leq 60$$
$$5x_1 + 2x_2 \leq 60$$
$$x_1, x_2 \geq 0.$$

If the constraints are satisfied, but the objective function is not maximized or minimized, we speak of a *feasible solution*. If the objective function is also maximized or minimized, we speak of an *optimal solution*. You can plot the constraints in an $x_1$-$x_2$ diagram, as shown in Figure 12.1 The shaded area is the feasible area. Now we move the objective function through the diagram until we have maximized its value. The optimal solution for LP problems is always located at the cross points of the constraints (such cross point is called a *vertex*). In our example we only look at two variables; you can imagine that for problems with substantially more variables a geometrical evaluation is impossible and that trying all possible vertex solutions is not very efficient.

A systematic method for finding the solution to an LP problem is the *simplex* method, which can be employed in an 8-step plan. (Step 1) First we have to

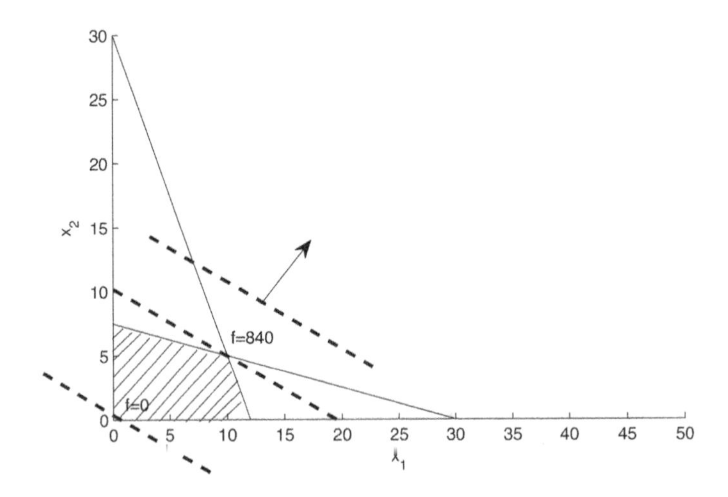


**FIGURE 12.1**
Geometrical representation of the LP problem

rewrite the model in its *normal form:*

$$\max z = f(x_1, x_2) = 40x_1 + 88x_2$$

$$s.t.$$

$$2x_1 + 8x_2 + x_3 = 60$$
$$5x_1 + 2x_2 + x_4 = 60$$
$$x_1, x_2, x_3, x_4 \geq 0.$$

$x_3$ and $x_4$ are called *slack variables.* They are the nonauxiliary variables introduced for the purpose of converting inequalities into equalities. (Step 2) We can write the problem in the normal form as the following augmented matrix:

$$T_0 = \left| \begin{array}{cccccc} z & x_1 & x_2 & x_3 & x_4 & b \\ 1 & -40 & -88 & 0 & 0 & 0 \\ 0 & 2 & 8 & 1 & 0 & 60 \\ 0 & 5 & 2 & 0 & 1 & 60 \end{array} \right| . \tag{12.1}$$

This matrix is called the (initial) simplex table. (Step 3) Each simplex table has two kinds of variables: the *basic variables* (columns having only one non-zero entry) and the *non-basic variables.* Every simplex table has a feasible solution. (Step 4) This solution can be obtained by setting the non-basic variables to zero, so for our example: $x_1 = 0$ and $x_2 = 0$. (Step 5) We can find values for $x_3$ and $x_4$ by using row 2 and row 3 of the simplex table:

$$2x_1 + 8x_2 + 1x_3 + 0x_4 = 60 \tag{12.2}$$
$$5x_1 + 2x_2 + 0x_3 + 1x_4 = 60. \tag{12.3}$$

From these two equations you can obtain: $x_3 = 60$ and $x_4 = 60$. These values for $x$ give us our first feasible solution with $z = 0$. Now we can calculate the optimal solution stepwise by pivoting in such a way that $z$ reaches a maximum. The big question now is how to choose your pivot equation. (Step 6) Select first the column with a negative entry; in our case this is column 2 ($-40$). This will be our pivot column. (Step 7) Now we divide the right-hand side $b$ by the pivot column and take the pivot that gives the smallest element. The first row does not count. So for row 2 we will have $60/2 = 30$ and for row 3 we will have $60/5 = 12$. (Step 8) Now we are going to eliminate all elements above row 3 with row operations. We can add to row 1 8 times row 3, and we can subtract from row 2 0.4 times row 3, resulting in the following table:

$$T_1 = \left| \begin{array}{cccccc} z & x_1 & x_2 & x_3 & x_4 & b \\ 1 & 0 & -72 & 0 & 8 & 480 \\ 0 & 0 & 7.2 & 1 & -0.4 & 36 \\ 0 & 5 & 2 & 0 & 1 & 60 \end{array} \right| . \tag{12.4}$$

With the new table we repeat Steps 3–8. First select the basic and non-basic variables. Set the non-basic variables to zero: $x_2 = 0$ and $x_4 = 0$, and calculate the values for the basic variables, which gives $x_1 = 15$ and $x_3 = 36$. These values will give us our second feasible solution with $z = 480$. Now select column 3 as the pivot column and identify which element has the largest quotient $36/7.2 = 5$ and $60/2 = 30$, ergo 7.2 will be the pivot element. Now we eliminate all elements above and below. We could add 10 times row 2 to row 1, and we could subtract $(2/7.2)$ times row 2 from row 3, resulting in the following table:

$$T_2 = \left| \begin{array}{cccccc} z & x_1 & x_2 & x_3 & x_4 & b \\ 1 & 0 & 0 & 10 & 4 & 840 \\ 0 & 0 & 7.2 & 1 & -0.4 & 36 \\ 0 & 5 & 0 & -1/36 & 1/0.9 & 50 \end{array} \right| . \tag{12.5}$$

There are no more negative entries, so this table contains the optimal solution, which is $x_1 = 50/5$, $x_2 = 36/7.2$, $x_3 = 0$, and $x_4 = 0$ with $z = 840$.

$z$ will increase as a result of the elimination of negative elements. It is noted that for a *minimization problem* all *positive elements* should be eliminated.

In MATLAB we can use the routing `linprog` to solve LP problems. For example, if we wish to solve the following problem:

$$\min f(x) = -5x_1 - 4x_2 - 6x_3$$

$$s.t.$$

$$x_1 - x_2 + x_3 \leq 20$$
$$3x_1 + 2x_2 + 4x_3 \leq 42$$
$$3x_1 + 2x_2 \leq 30$$
$$x_1, x_2, x_3 \geq 0,$$

we can define in MATLAB:

```
>>f = [-5; -4; -6];
>>A = [1 -1 1; 2 2 4; 3 2 0];
>>b = [20; 42; 30];
>>lb = zeros(3,1);
>>[x,fval,exitflag,output,lambda] = linprog(f,A,b,[],[],lb);
```

which will give you

```
x = 0.00 15.00 3.00.
```

## 12.3 Nonlinear programming

In nonlinear programming (NLP), the objective function $f(x)$ and/or the constraints $g(x)$ and $h(x)$ can be nonlinear functions, for example:

$$\max f(x)$$

$$s.t.$$

$$g(x) \leq 0$$
$$h(x) \leq 0$$
$$x \geq 0.$$

NLP problems can have a "free optimum" (or unconstrained optimum), in contrast to LP problems. There are several ways to solve NLPs, but in this chapter we will only demonstrate the *Lagrange multiplier method.* The Lagrange multiplier method is founded on the *Karush–Kuhn–Tucker conditions* for optimality. The derivation and proof are extensive, but the basic idea is as follows. Given the NLP problem above, we can define a so-called Lagrangian function

$$L = f(x) + \lambda g(x) + uh(x), \tag{12.6}$$

or for systems:

$$L = f(x) + \sum_i \lambda_i g_i(x) + \sum_j u_j h_j(x), \tag{12.7}$$

where $\lambda_i$ and $u_j$ are called the *Lagrange multipliers.* The optimality conditions state that for an optimal solution the following should hold:

$$\nabla L = 0. \tag{12.8}$$

For example, we would like to solve

$$\max f(x_1, x_2) = 5x_1^2 + 3x_2^2$$
$$s.t.$$
$$2x_1 + x_2 - 5 \leq 0$$
$$x \geq 0.$$

We first define the Lagrangian function:

$$L = (5x_1^2 + 3x_2^2) + \lambda(2x_1 + x_2 - 5). \tag{12.9}$$

To find an optimal solution we now have to set the partial derivatives of the Lagrangian function to zero:

$$\frac{\partial L}{\partial x_1} = 10x_1 + 2\lambda = 0 \tag{12.10}$$
$$\frac{\partial L}{\partial x_2} = 6x_2 + \lambda = 0 \tag{12.11}$$
$$\frac{\partial L}{\partial \lambda} = 2x_1 + x_2 - 5 = 0. \tag{12.12}$$

These are actually three equations with three unknowns, which can be found easily: $\lambda = 150/17$, $x_1 = 30/16$, and $x_2 = 25/17$.

To solve NLP problems in MATLAB, we can use the function `fmincon` (for problems with constraints) and `fminsearch` (for unconstrained problems).

Suppose we want to solve the following problem:

$$\min f(x_1, x_2, x_3) = -x_1 x_2 x_3$$
$$s.t.$$
$$0 \leq x_1 + 2x_2 + 2x_3 \leq 72.$$

We should first write a function containing the objective:

```
function f = myfun(x)
f = - x(1) * x(2) * x(3);
```

We also should define the constraints as

```
>>A = [-1 -2 -2; 1 2 2];
>>b = [0 72];
```

and we should supply a guess value:

```
>>x0 = [10; 10; 10];
```

Now we may find the optimum by:

```
>>[x,fval] = fmincon(@myfun,x0,A,b);
```

which will yield:

```
x = 24.00 12.00 12.00
```

There are some tips for solving NLP problems. Try to avoid nonlinearity as much as possible. It is better to have nonlinearities in the objective function than in the constraints. It is also better to have inequality constraints than equality constraints. A good starting guess is very important. And, do not blame the solver if you do not find a solution; take a critical look at the problem formulation.

## 12.4 Integer programming

Integer problems occur a lot in chemical engineering, especially in design, scheduling, and planning of process systems. Often, so-called mixed integer problems (MIPs) have to be solved, where we have a combination of discrete and continuous variables. Such problems may be nonlinear, e.g., the MINLP (mixed integer nonlinear program) or in a specific case quadratic (MIQP). However, in integer programming (IP) all optimization variables are discrete, and the problem would look like this:

$$\max f(x)$$
$$s.t.$$
$$g(x) \leq 0$$

$$h(x) \leq 0$$
$$x \in \mathbb{N}.$$

There exist several algorithms to solve this type of problem but many of them are based on the *branch and bound* algorithm. The algorithm is best explained with an example.

Suppose we want to solve the following problem:

$$\max z = 8x_1 + 11x_2 + 6x_3 + 4x_4$$
$$s.t.$$
$$5x_1 + 7x_2 + 4x_3 + 3x_4 \leq 14$$
$$x \in \{0, 1\}.$$

It is noted that $x$ is a special type of integer variable. Because it can only assume 0 or 1, it is called a *binary variable*, a yes or no decision.

The first step in the branch and bound algorithm is to solve the *relaxed* problem, that is, we remove the integrality constraint $x \in \{0, 1\}$ from the problem and solve it as a continuous problem. We could solve the relaxed problem with, for example, the simplex method, and find $x_1 = 1$, $x_2 = 1$, $x_3 = 1/2$, and $x_4 = 0$, with the objective value being $z = 22$. This value for $z$ actually forms an *upper bound* to the problem. The problem occurs with $x_3$, which should, of course, have been an integer. We can now solve two new problems: in one version we will add a constraint to the problem with $x_3 = 0$ and in the second one we add a constraint $x_3 = 1$. So we actually *branch* on the original problem, as is shown in Figure 12.2. Now we can select an active subproblem from the branches. Say we continue with the right branch of the tree. We have

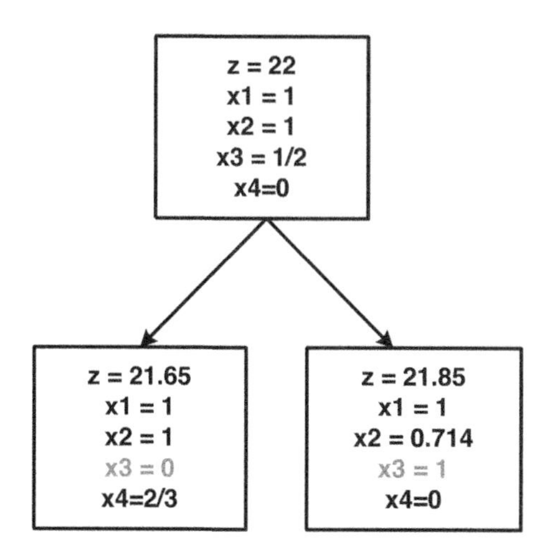


**FIGURE 12.2**
Branching the relaxed problem

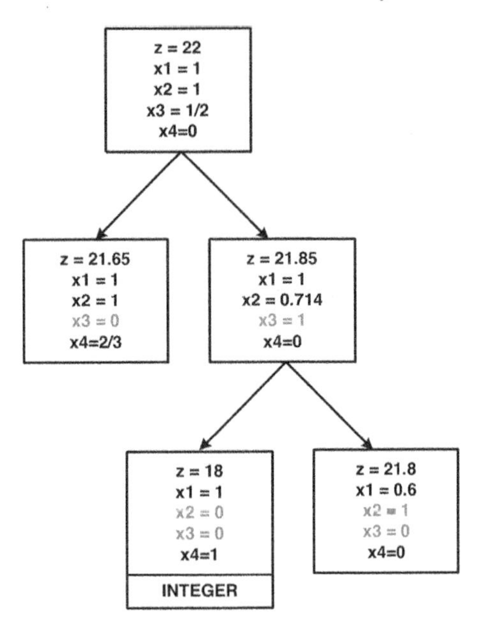


**FIGURE 12.3**
Second branch of the relaxed problem

added a constraint to the problem $x_3 = 1$ and now find an optimal solution $x_1 = 1$, $x_2 = 0.714$, and $x_4 = 0$ with an objective value of $z = 21.85$. We should now branch on $x_2$ as it is not an integer. So we add two subproblems: one in which we solve for $x_2 = 0$ and one in which we solve for $x_2 = 1$, as shown in Figure 12.3. In the left branch we have already found an integer solution, but it might not be the optimal one. So we continue branching on the right part of the tree, where we have found that $x_1$ is not an integer. So we add two more subproblems, as shown in Figure 12.4. The right branch cannot be solved, as it has become an infeasible problem. The left branch again results in an integer solution, but with a higher objective value than the previously calculated one $z = 21$. This value for $z$ is actually the maximum value that can be obtained, ergo we have found the optimum. For this reason it is of no use to look in the other branch of the tree, and we can *fathom* that line of our search.

It should be noted that the branch and bound algorithm only works if the objective function is convex.

## 12.5 Summary

Optimization is concerned with finding the extremum (minimum or maximum) of an objective function. The optimization variables can be constrained

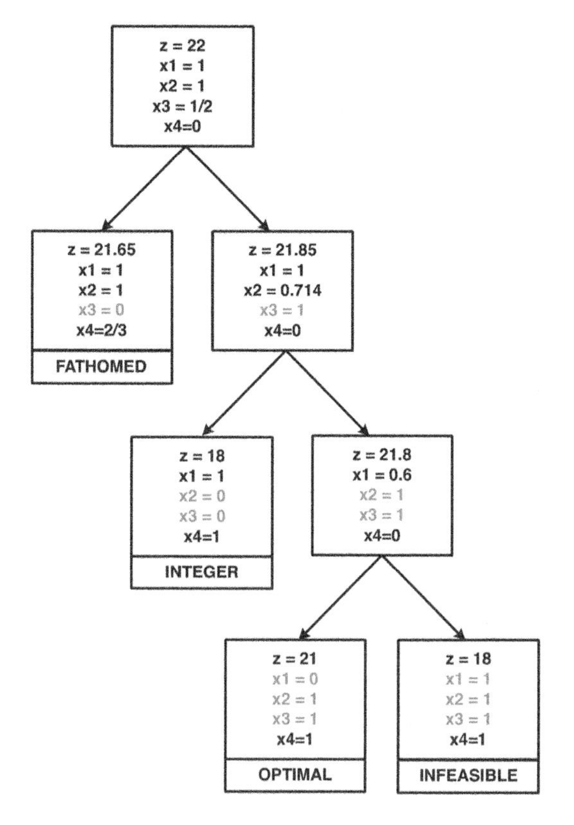


**FIGURE 12.4**
Third branch of the relaxed problem

by equability or inequality constraints. We approached LP problems with the simplex algorithm, NLP problems with the Lagrange multiplier method, and IP with the branch and bound algorithm.

## 12.6 Exercises

### Exercise 1

In this exercise we are going to solve a large-scale LP problem of the following form:

$$\begin{aligned} \min f^T x \\ s.t. \\ & A_{eq}x = b_{eq} \\ & x_{lb} \leq x \leq x_{ub}. \end{aligned}$$

You can load the matrices and vectors $A_{eq}$, $b_{eq}$, $f$, $x_{lb}$, and $x_{ub}$ into the MATLAB workspace with `load densecolumns`.

The problem is `densecolumns.mat` has 1677 variables and 627 equalities with lower bounds on all the variables, and upper bounds on 399 of the variables. The equality matrix $A_{eq}$ has dense columns among its first 25 columns, which is easy to see with a spy plot. Type `spy(Aeq)`.

We are going to use `linprog` to solve the system. Type the following code:

```
[x,fval,exitflag,output] = ...
linprog(f,[],[],Aeq,beq,lb,ub,[],optimset('Display','iter'));
```

### Exercise 2

In this exercise we are going to use `fmincon` to solve the following optimization problem:

$$\begin{aligned} \min f(x_1, x_2) &= \exp(x_1)(4x_1^2 + 2x_2^2 + 4x_1x_2 + 2x_2 + 1) \\ s.t. & \\ & x_1^2 + x_2 = 1 \\ & -10 \leq x_1x_2. \end{aligned}$$

The first step is to write a function for the objective function, such as:

```
function f = objfunex1(x)
f = exp(x(1))*(4*x(1)^2+2*x(2)^2+4*x(1)*x(2)+2*x(2)+1);
```

The second step is to write a function for the (in)equality constraints; pay attention to the formulation of the constraints!

```
function [c, ceq] = confuneqex1(x)
c = -x(1)*x(2) - 10;
ceq = x(1) ^2 + x(2) - 1;
```

The third step is to apply the solver. First define the optimization options:

```
options=optimset('Display','iter','Jacobian','on');
```

and give a set of starting guesses for optimization:

```
x0 = [-1,1];
```

```
[x,fval] = fmincon(@objfunex1,x0,[],[],[],[],[],[],...
@confuneqex1,options)
```

Interpret the data. Are the constraints satisfied with the optimal values for $x_1$ and $x_2$?

### Exercise 3

We have two plants and three markets to which we would like to ship products. There are several ways to do this, but we want to do it the most efficient way.

The problem in the chapter was given as:

$$\min \sum_i \sum_j c_{ij} x_{ij}$$

$$s.t.$$

$$\sum_i x_{ij} \leq a_i$$

$$\sum_j x_{ij} \leq b_j.$$

You can open the files `TRANSP1.MOD` and `TRANP1.DAT` in a text editor to see how the problem can be formulated in AMPL. AMPL always requires a model file and a data file.

After you download AMPL from the web and extract the program, you execute `sw.exe`. Then type `ampl`.

First open the model file by:

```
model c:\examples \transp1.mod;
```

Don't forget the semicolon (;).

Next, open the data file with:

```
data c:\examples \transp1.dat;
```

We will solve the model with a solver called MINOS:

```
option solver minos;
solve;
```

The data is recorded under NoUnits, NoUnits.rc, and total_cost. You can display the optimized values by typing:

```
display NoUnits, NoUnits.rc;
display total_cost;
```

# 13

# *Basics of MATLAB*

## 13.1 Introduction

This instruction chapter is intended to familiarize the beginning user of MATLAB with its basic features. MATLAB release 2010a was used in this book.

The official MATLAB website is also very useful in exploring the capabilities of MATLAB. The reader can access `http://www.mathworks.com`. In addition, it should be mentioned that by typing `help` or `doc` followed by the keyword of interest, a clear explanation is given of MATLAB's functionality, including many worked-out examples. For a complete MATLAB instruction, the reader is referred to the textbook by Hahn and Valentine [24].

## 13.2 The MATLAB user interface

When you open MATLAB, the screen depicted in Figure 13.1 appears. The screen exists of three basic subscreens. On the left you will find the "work space" and the "command history." On the right side you find the "command window." In the work space all variables and objects that you construct are stored and in the command history, you will find an overview of the commands that you executed (with arrows up and down you can scroll and execute previous commands very rapidly). If you want to save or open a file, it is always useful to change the "current directory" to the correct path.

You can save variables, data, and objects in the work space by typing in the command prompt `>>save filename` and you can open files with `>>open filename`.

You communicate with MATLAB via the command prompt. It is noted however that you do not really have to type each command separately in the command window; you can also automate the commands by saving them in a file, a so-called "m-file" or "script." You can make an m-file by choosing in the file menu (left) the option `new/m-file`. A text editor opens and you can

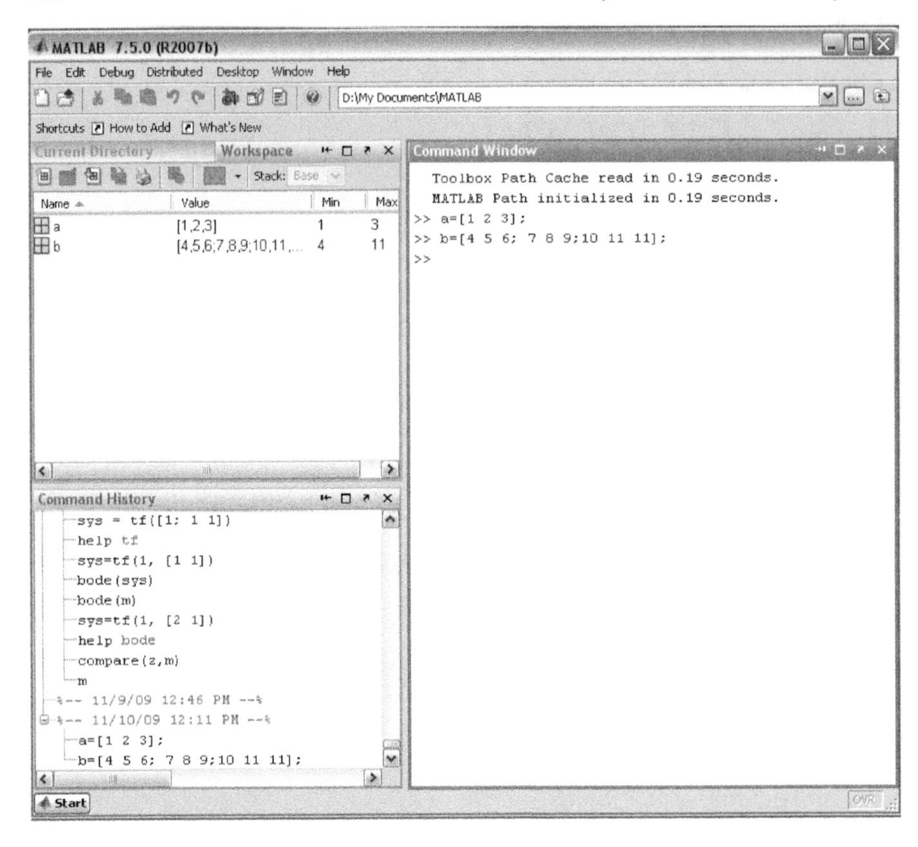


**FIGURE 13.1**
MATLAB screenshot

type the MATLAB commands in the text editor, save the script, and then you can execute it by choosing the option `debug/run` or by pressing the F5 button. It is also possible to call the script via the command prompt. Just type the filename and it will execute. You should be aware that commands and files in MATLAB are case sensitive. MATLAB can accept filenames of 255 characters, but the first character of a filename cannot be a number or a symbol. Use commonsense when giving names to files!

## 13.3 The array structure

MATLAB stands for *Matrix Laboratory.* Matrix and vector calculations are an intrinsic part of MATLAB. Defining variables in MATLAB is based on the "array structure." You could, for example, define a vector $b$ and store it in the work space by typing:

```
>> b = [ 1 2 3 4 5];
```

Or, if you would like to define a matrix A:

```
>> A = [1 2 3; 2 3 4; 3 4 5]
```

If you would like to use a specific element of this matrix for further calculations, for example, the third element in the second column, you could access it by typing:

```
>> A(2,3)
```

If you would like to access the entire third column, you could use the colon:

```
>> A(:,3)
```

Or if you would like to access the entire second row:

```
>> A(2,:)
```

You could also use the `end` command, for example, if you wanted to access the second-to-the-last element of the first column:

```
>> A(2:end,1)
```

Now if you want to define a vector $x$ of 100 elements over the domain 0 to 10 you do not need to type out each of the elements individually. You could create the vector $x$ in MATLAB by typing:

```
>> x = 0:0.1:10
```

In other words: `x = startvalue : stepsize : finalvalue`. Here it is also noted that you could transpose matrices and vectors with the quote ('), for example:

```
>> x'
```

MATLAB has really powerful tools for matrix calculations. You could invert a matrix by typing:

```
>> \A
```

but `inv(A)` or `A^-1` would also work. Just for your information, you could also calculate, for example, the eigenvalues and eigenvectors (`eig(A)`), the determinant (`det(A)`), and the rank (`rank(A)`) of matrices very easily.

## 13.4 Basic calculations

After defining a vector $x$, from 0 to 10 in steps of 0.1, perhaps you would like to calculate $y$-values that are linked to $x$ according to

$$y(x) = 2\sin(5x). \tag{13.1}$$

| Standard | MATLAB | Standard | MATLAB |
|---|---|---|---|
| $\sin(x)$ | sin(x) | $\sqrt(x)$ | sqrt(x) |
| $\cos(x)$ | cos(x) | $e^x$ | exp(x) |
| $\tan(x)$ | tan(x) | $\ln(x)$ | log(x) |
| $\sin^{-1}(x)$ | asin(x) | $^10\log(x)$ | log10(x) |
| $\cos^{-1}(x)$ | acos(x) | $\|\|x\|\|$ | abs(x) |
| $\tan^{-1}(x)$ | atan(x) | $sign(x)$ | sign(x) |

**TABLE 13.1**
Basic MATLAB functions

| Standard | MATLAB |
|---|---|
| $e$ | exp(1) |
| $\pi$ | pi |
| $j$ | i or j |
| $\infty$ | inf or Inf |

**TABLE 13.2**
Basic MATLAB constants

In MATLAB you could easily calculate the values for $y$ on the basis of $x$:

```
>> y = 2*sin(5*x)
```

Similarly, you could compute:

$$y(x) = 2x^2 + 4x - 5 \tag{13.2}$$

by feeding to the command window:

```
>> y = 2*x ^2 + 4*x - 5
```

But, MATLAB responds with an error:

```
??? Error using ==> mpower Matrix must be square
```

Because MATLAB is a vector- and matrix-based computation tool, it intends to do matrix multiplications and divisions, which are basically dot products, inner products, or matrix inversions, etc. So scalar types of calculations cannot be performed so easily. To make sure that MATLAB will perform the computation, you have to add a dot to the operation (sometimes referred to as the *dot operator*), like this:

```
>> y = x. ^2 + 4.*x -5
```

In Tables 13.1, 13.2, and 13.3 you will find a list with commonly performed mathematical operations and the MATLAB expressions that go with them.

| Standard | MATLAB |
|---|---|
| $a + b$ | `a + b` |
| $a - b$ | `a - b` |
| $ab$ | `a * b` |
| $\frac{a}{b}$ | `a ^b` |
| $a^b$ | `a \b` |

**TABLE 13.3**
Basic MATLAB operations

## 13.5 Plotting

We can make a plot of $x$ and $y$ easily in MATLAB by typing:

```
>> plot(x,y)
```

Or if you like dots more than lines, type the following:

```
>> plot(x,y,'.')
```

You could also manipulate the color of the graph, for example, using green dots:

```
>> plot(x,y,'g.')
```

There are many other possibilities for color (`k,b,r,...`), symbols (`.,` `+,*,` `^,` `o,` `...`), and lines (`-,` `--,` `-.,...`).

You could give a title to your plot by:

```
>> title('y as function of x')
```

and you could add titles to the axis with:

```
>> xlabel('x-values')
>> ylabel('y-values')
```

You could change the scaling of the axis by, for example:

```
>> axis([ 0 10 0 100])
```

where the syntax is `[minimal-x, maximal-x, minimal-y, maximal-y]`. Suppose that now you have $y_1(x)$ and $y_2(x)$:

```
y1 = 15*sin(5*x); y2 = 2.^2 + 4.*x - 5
```

and you want to plot both function in one graph:

```
plot(x,y1,'ro',x,y2,'g*')
```

When you want to have two independent graphs, you could use subplot:

```
>> subplot(1,2,1); plot(x,y1,'ro')
>> subplot(1,2,2); plot(x,y2,'g*')
```

You can save figures, copy them, and then use them, for example, in MS Word.

## 13.6 Reading and writing data

As discussed earlier, basic reading and writing of data can be done with the MATLAB `load` and `save` functions. But data from text files or Excel files can also be pasted to the work space directly from the menu bar: `edit/paste to workspace`. In addition, there are builds in MATLAB functions that you can use to import data from, for example, Excel sheets. With the command `xlsread` you can import data, and with `xlswrite` you can write MATLAB data to Excel sheets. These commands are very useful when many data files are to be examined.

## 13.7 Functions and m-files

We mentioned earlier that MATLAB commands can be automated. For example, you could make a script or m-file where you define the vector $x$, calculate $y$, and then plot $x$ versus $y$:

```
x=0:0.1:10
y=15*sin(2*x)
plot(x,y,'ro')
```

You can open the script editor in MATLAB and write the above commands in a script. Next you can save this file as "makeplot.m." Now you can execute the file by pressing F5 or by typing `makeplot` in the command window. In this way you do not need to retype all three commands, every time in the command window, in case you want to make small changes.

Alternatively, you could write a MATLAB function. A function performs tasks on the basis of input information that you can supply to the script in an interactive way.

We could, for instance, make a function that uses a given domain and a number of points that we want for $x$, then compute $y$ values and ultimately plot the result:

```
function makeplot(minx,maxx,numberofpoints)
```

```
Deltax = maxx-minx)/numberofpoints;
x=minx:Deltax:maxx;
y=15*sin(2*x);
plot(x,y,'ro');
```

The semicolon behind each command ensures that the result of each command is computed, but not written or displayed, in the command window (this is very handy in case you perform repetitive operations, which would slow down your calculations if the result is written on your screen each time).

Earlier we defined a vector $x$ and calculated $y$ by typing the equations in the command window. However, you could also define a "function handle":

```
function y = myfun(x)
y = 15*sin(2*x);
```

Save this function as `myfun`. Now, when you type the following in the command window:

```
>> y = myfun(10)
```

the command window will give you the calculated $y$ value for $x = 10$. Similarly, if you type

```
>> y = myfun(1:10)
```

MATLAB will not return an array with values of $y$ because $x$ is 1 to 10. Storing mathematical expressions in function handles is very useful, for example, in solving linear, nonlinear, and differential equations.

Try always to keep your scripts organized. With the percent sign, you can also add comment lines to your program. Such comment lines are not part of the code, but they make sure that you remember, or your fellow team members remember, what you actually meant or were doing when writing the program:

```
plot(x,y,'ro') % plotting x versus y with red circles
```

## 13.8 Repetitive operations

Sometimes you need to repeat calculation steps frequently with different input data, or you would like to use results of previous calculations for new calculations. Such calculations are called *repetitive operations* or *iterations*. In MATLAB you can do iterations in two ways, with the so-called "for-loop" and the "while-loop."

Here is a small example. Suppose you want to compute a series of *Fibonacci* numbers. Fibonacci was an Italian scientist from the Renaissance. He developed iterative schemes to calculate the *golden mean* number, but he also studied natural phenomena, such as the reproduction rates of bunnies. He found that a certain series of numbers could be found in nature quite often: 1, 1, 2, 3, 5, 8, 13, 21,... This series can actually be calculated from a recurrent relationship:

$$y_k = y_{k-1} + y_{k-2}. \tag{13.3}$$

A new Fibonacci number is the sum of the two previous numbers, where $y_1 = 1$ and $y_2 = 1$. If you divide two sequel Fibonacci numbers, you will get an approximation of the golden mean, $\phi$.

In a for-loop we can calculate a series of Fibonacci numbers. Open the script editor, and save the following commands in a file:

```
y(1) = 1
y(2) = 1
for i = 3:10
 y(i) = y(i-1) + y(i-2);
 phi = y(i)/y(i-1);
end
```

You can also add so-called "conditional statements" to a loop. In the example below, we will construct a step function, where a function $f$ will have a value of $-1$ for the time domain $t = -10...0$, and where $f$ will have a value of 0 for $t = 0...10$:

```
t = -10:0.1:10;
n=max(size(t));
f = zeros(1,n);
for i=1:n
 if t(i) < 0;
 f(i) = -1;
 elseif t(i) == 0;
 f(i) = 0;
 elseif t(i) > 0;
 f(i) = 1;
 end
end
plot(t,f)
```

MATLAB also uses the while-loop. In a while-loop, an operation is repeated until a certain end criterion is reached. In Chapter 6 there is a demonstration of the while-loop in Newton's method for solving nonlinear equations.

# 14

# *Numerical methods in Excel*

## 14.1 Introduction

This instruction chapter is intended to familiarize the user of Excel with basic numerical methods in Excel. Excel 2007 was used in this book.

Microsoft Excel is a commonly used spreadsheet program and has developed into a powerful software that can be used to solve many science and engineering problems. In contrast to MATLAB, Excel is most probably available in almost all PCs and it appears that its usage is increasing. It is also noted that the OpenOffice suite is a shareware version of Microsoft Office which also contains a spreadsheet similar to Excel, called *calc*. Excel is also used in education and it seems obvious that we also devote some attention to the basics of Excel and how it can be used to solve numerical method problems.

## 14.2 Basic functions in Excel

The reader is probably familiar with MS Excel. For that reason this section is only used to tabulate the most useful Excel functions. You will find the basic Excel functions in Table 14.1 and the array formulae in Table 14.2. Excel can compute matrix inverses, which opens the door to solving systems of linear equations. Note, however, that Excel's array tools are not suitable for very large systems.

## 14.3 The Excel solver

The Excel solver function is a nice and very useful tool within Excel. If you have not installed the solver add-in, you can consult Excel Help to make sure

| Standard | Excel | Standard | Excel |
|---|---|---|---|
| $\sin(x)$ | =SIN(x) | $\sqrt(x)$ | =SQRT(x) |
| $\cos(x)$ | =COS(x) | $e^x$ | =EXP(x) |
| $\tan(x)$ | =TAN(x) | $\ln(x)$ | =LN(x) |
| $\sin^{-1}(x)$ | =ATAN(x) | $^{1}0\log(x)$ | =LOG(x) |
| $\cos^{-1}(x)$ | =ACOS(x) | $\|x\|$ | =ABS(x) |
| $\tan^{-1}(x)$ | =ATAN(x) | | |

**TABLE 14.1**
Basic Excel functions

| Standard | Excel |
|---|---|
| summation | =SUM(x) |
| average | =AVERAGE(x) |
| standard deviation | =STDEV(x) |
| matrix inverse | =MINVERSE(x) |
| matrix multiplication | =MMULT(x) |
| matrix transpose | =TRANSPOSE(x) |
| random number 0...1 | =RAND() |

**TABLE 14.2**
Basic Excel array formulae

the solver is available. The solver add-in allows you to solve optimization problems, i.e., to find minima or maxima for mathematical expressions. We will illustrate how the Excel solver works with a very simple example.

Suppose you would like to find the maximum of a parabola

$$f(x) = -2x^2 + 5x - 3. \tag{14.1}$$

You could enter a value in cell A2, say =3. In cell B2 you can now enter the formula to calculate $f(x)$. Enter: `=-2*A2^2 + 5*A2-3`. Of course we now want to find the value for $x$ that is at the maximum of $f$. Under the "data" tab, in the right upper corner of the Excel menu you should be able to find the Solver add-in. If you click this option, a small window will open, as shown in Figure 14.1. We now set the "target cell" to B2 and fill in under "By changing" cell, A2. In this case we want to maximize B2, so we select under "Equal to" the *max* option. Now we click "Solve" and Excel will compute for which value of $x$, $f(x)$ reaches the maximum, in this case $x = 1.25$ and $f = 0.125$.

You can understand that this solver option opens the possibility for solving many engineering problems, including curve-fitting exercises.

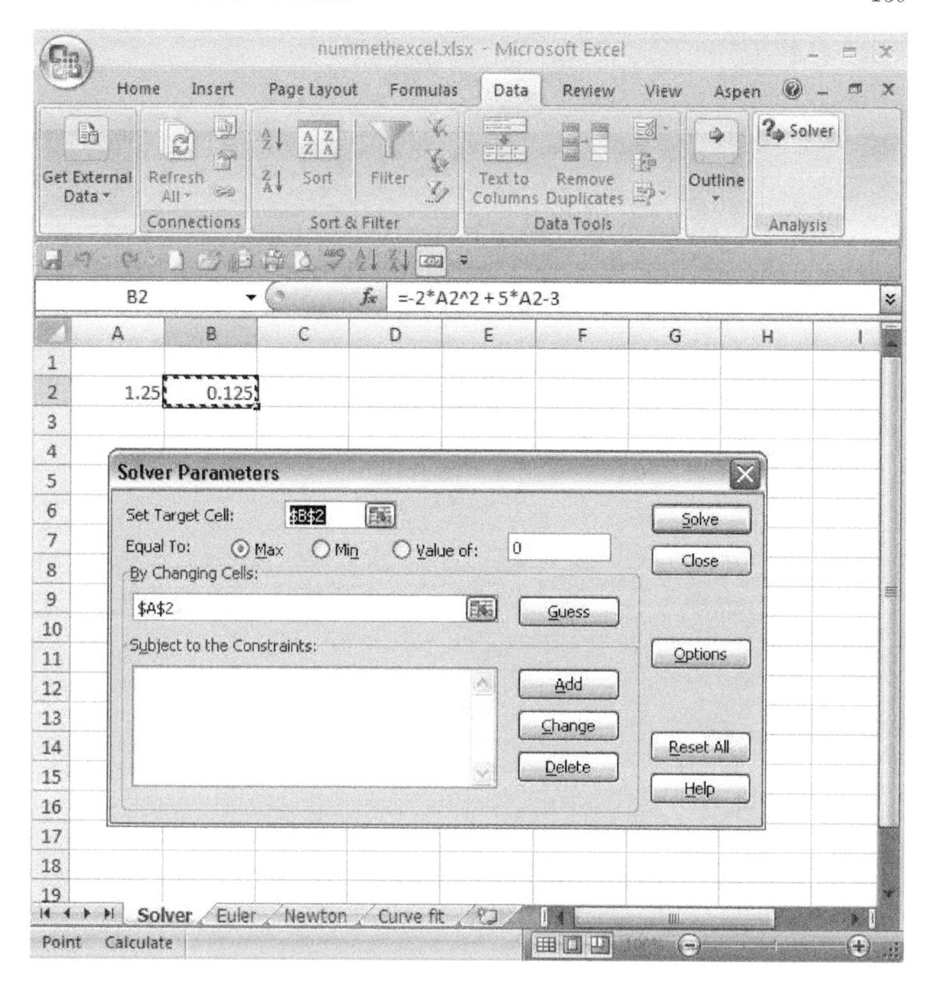


**FIGURE 14.1**
Excel screenshot for the solver add-in

## 14.4 Solving nonlinear equations in Excel

In this section we are going to implement Newton's method in Excel. With Newton's method we can find the roots of nonlinear equations. You may recall that the basic equation of Newton's method is

$$\Delta x = \frac{-f(x)}{f'(x)}, \tag{14.2}$$

where $\Delta x$ is the Newton step that we take from a starting point $x_n$ to find the root of a function $f(x)$. We further recall that Newton's method is an

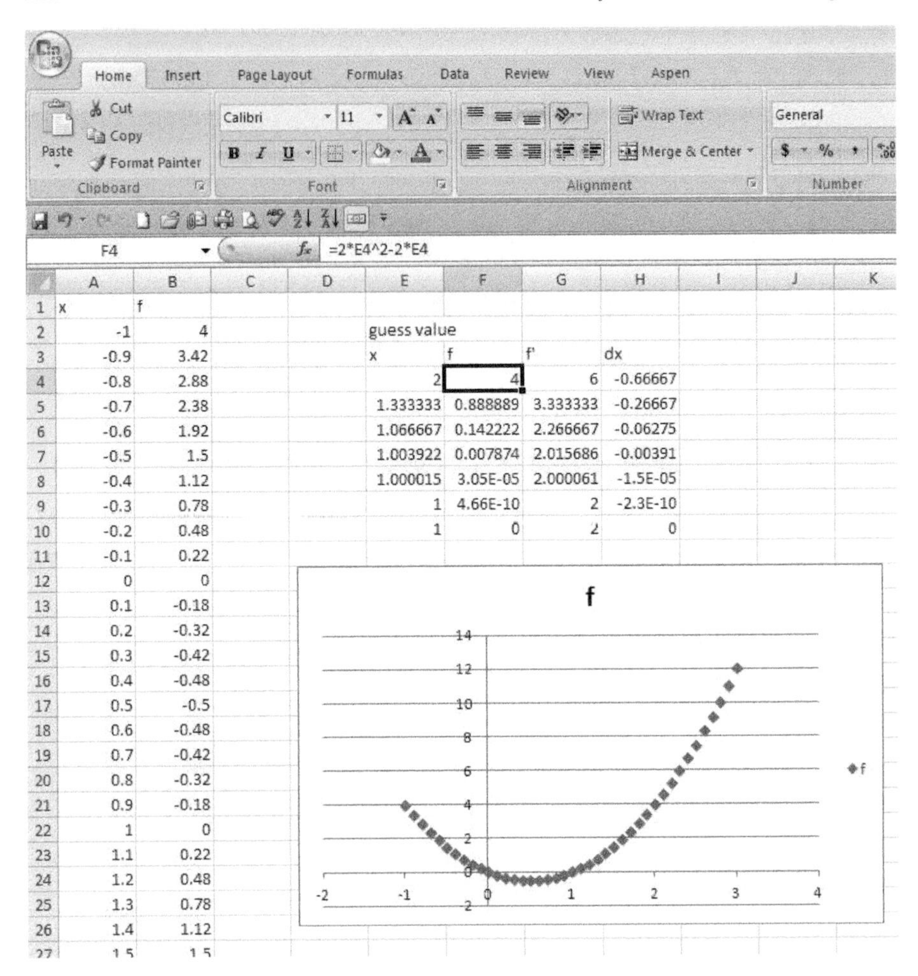


**FIGURE 14.2**
Excel screenshot for Newton's method

iterative procedure and that a new estimate should be obtained from

$$x_{n+1} = x_n + \Delta x. \tag{14.3}$$

This iteration should proceed until we are sufficiently close to $f(x) = 0$, in other words, $\Delta x = 0$.

Let us use Excel to find the root of a simple parabolic equation:

$$f(x) = 2x^2 - 2x. \tag{14.4}$$

In Figure 14.2 a screen shot of the Excel sheet is given. In cell E4 an initial guess for $x$ is provided. In cell F4 you can enter the formula =2*E4^2-2*E4.

This calculates what $f(x)$ should be for $x = 2$. In cell G6 we will evaluate the derivative of $f(x)4$ at $x = 2$. You can type `=4*E4-2`. Subsequently we can use Newton's equation to calculate the step we shall take until the next iteration. Type in cell H4 `=-F4/G4`. Now we can obtain a new estimate for $x$. Enter in cell E5 `=E4+H4`, which is the previous value of $x$ plus $\Delta x$. With this new estimate of $x$ we can compute new values for $f(x)$, $f'(x)$, and $\Delta x$, as shown in the screen shot.

After 5 iterations it is noted that convergence has been reached, $\Delta x = 0$ and $x = 1$ does not change anymore. This is the root of Equation 14.3. If a different start value would have been provided, for example, $x = -1$, we would have found the other root at $x = 0$.

## 14.5 Differentiation in Excel

In a similar fashion, we can also solve differential equations with Excel. We will have a look at the implicit Euler method, which was discussed in previous chapters. Suppose we have an initial value problem

$$\frac{dx}{dt} = f(x,t) = kx, \tag{14.5}$$

where $k = -0.5$ and we further know that $x(0) = 1$. Using Euler's methods we can march a numerical solution forward in time with a Taylor series:

$$x(t + \delta t) = x(t) + f(x,t)\delta t. \tag{14.6}$$

Let us now use Excel to solve our ODE numerically with Euler's method. We will use a $\delta t = 0.1$. In Figure 14.3 a screen shot is given. In the first column we have defined values for the time. In the second column we will calculate values for $x$. We start off with the initial value for $x$ at $t = 0$. Enter in cell B3 `=1`.

Now we can continue with the implementation of Euler's method. In cell B4 we now enter `=(1-0.5*0.1)*B3`. We can then copy cell B4 to cells B5 and further. You can test for yourself how close this numerical solution is to the analytical expression $x(t) = x_0 e^{kt}$.

## 14.6 Curve fitting in Excel

In Figure 14.4 a screen shot is given for a curve-fitting example in Excel. Curve fitting is fairly easy and user friendly in Excel. As can be seen from

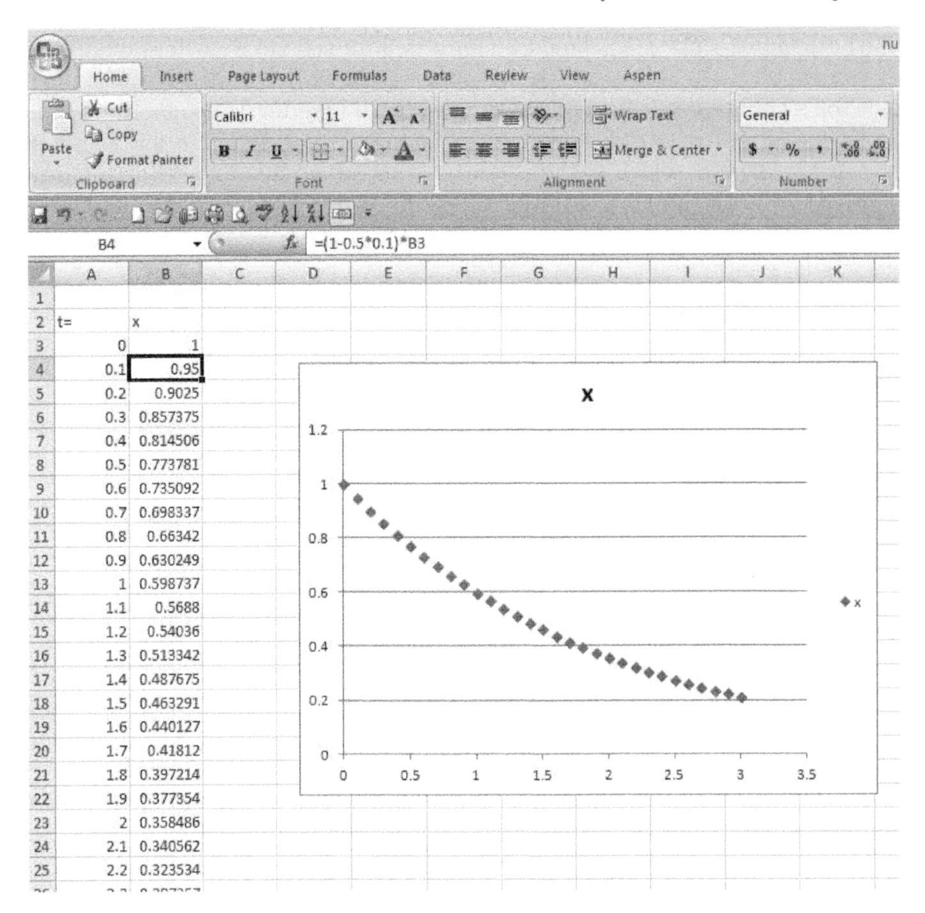


**FIGURE 14.3**
Excel screenshot for Euler's method

the data in Figure 14.4 we are dealing with some kind of polynomial data. If you right-click the data, a pop-up opens with the possibility to "add a trend line." If you select this option, a window opens with several options for fitting, linear, polynomial, exponential, etc. It is also possible to display the fitted parameters and the regression coefficient in the figure.

Now that you know how to use the solver option in Excel, you can also develop your own fitting tool.

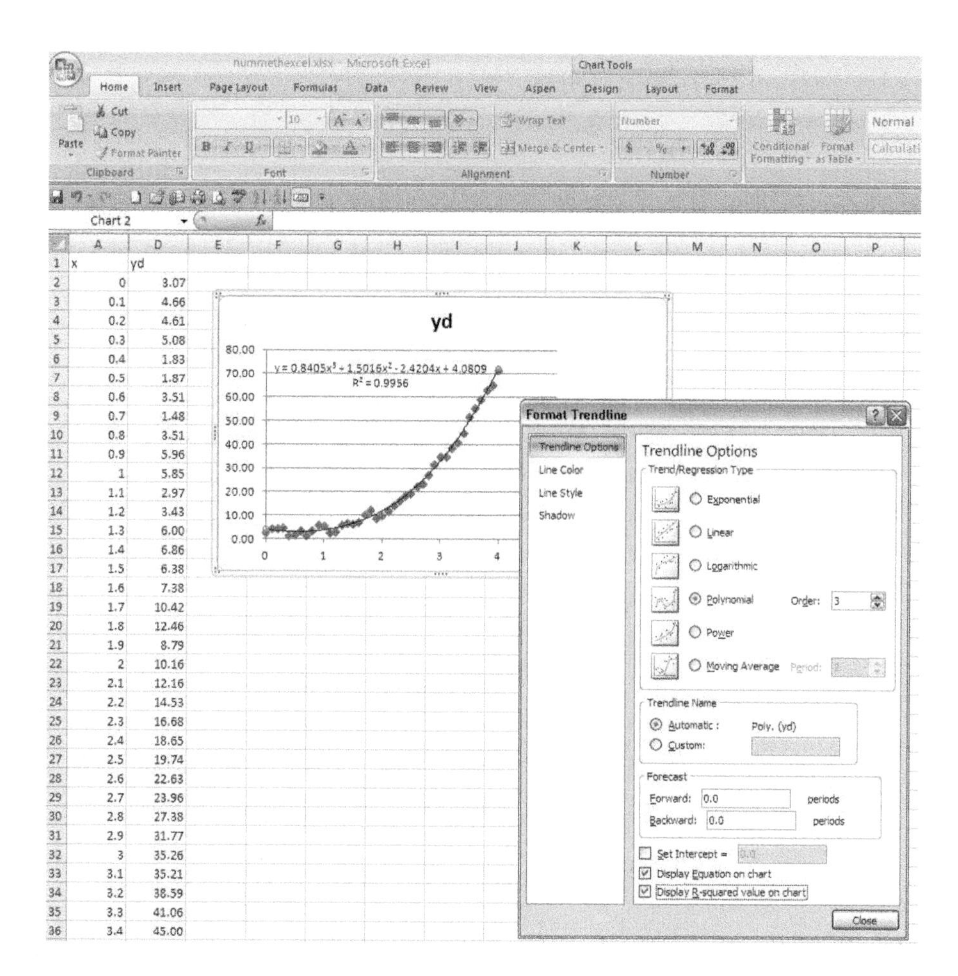


**FIGURE 14.4**
Excel screenshot for curve fitting

# 15

# *Case studies*

## 15.1 Introduction

In this chapter several examples are given for problems that the chemical engineer needs to solve using numerical methods.

## 15.2 Modeling a separation system

For the separation system of Figure 15.1, we know the inlet mass flow rate (in kilograms per hour) and the mass fractions of each species in the inlet (stream 1) and each outlet (streams 2, 4, and 5). Compute the mass flow rates of each of the outlet streams.

We will use the notation where $F_i$ is the mass flow rate of stream $i$ and $w_{i,j}$ is the mass fraction of species $j$ in stream $i$. The unknowns are defined as

$$x_1 = F_2, x_2 = F_4, x_3 = F_5.$$

Set up the balance for the total mass flow rate, the mass flow rate of species 1, and the mass flow rate of species 2 of the separation system and compute the mass flow rates of each of the outlet streams using (a) Gaussian elimination (by hand), (b) Gaussian elimination (with MATLAB) and (c) two alternatives in MATLAB. Compare the different methods.

The total mass flow rate is

$$F_2 + F_4 + F_5 = F_1. \tag{15.1}$$

We can set up the mass flow rate of species 1 with

$$w_{1,2}F_2 + w_{1,4}F_4 + w_{1,5}F_5 = w_{1,1}F_1 \tag{15.2}$$

and the mass flow rate of species 2 is

$$w_{2,2}F_2 + w_{2,4}F_4 + w_{2,5}F_5 = w_{2,1}F_1, \tag{15.3}$$

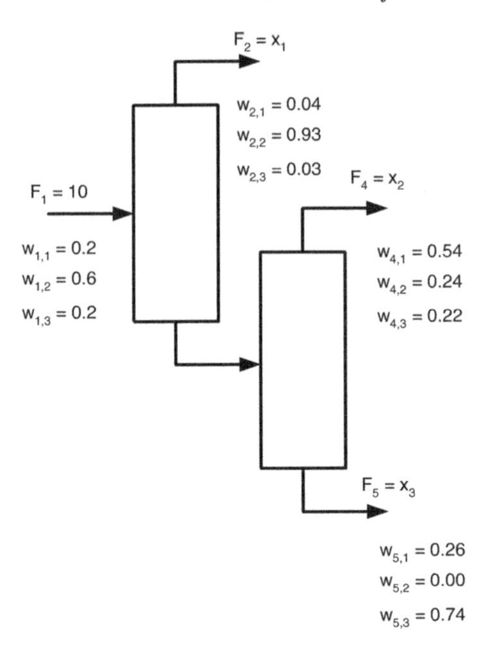


**FIGURE 15.1**
Process diagram for the separation system

which yield, after rewriting, a set of three linear equations:

$$\begin{bmatrix} 1.00 & 1.00 & 1.00 \\ 0.04 & 0.54 & 0.26 \\ 0.93 & 0.24 & 0.00 \end{bmatrix} \begin{bmatrix} F_2 \\ F_4 \\ F_5 \end{bmatrix} = \begin{bmatrix} 10 \\ 2 \\ 6 \end{bmatrix}. \tag{15.4}$$

With Gaussian elimination in MATLAB, this simple system of three equations with three unknowns can be solved easily:

```
>>A = [ 1.00 1.00 1.00; 0.04 0.54 0.26; 0.93 0.24 0.00];
>>b = [ 10 2 6];
>>x = A \b;
```

which will give as a result $F_2 = 5.82$, $F_4 = 2.43$, and $F_5 = 1.74$.

## 15.3 Modeling a chemical reactor system

A chemical reaction takes place in a series of four continuous stirred tank reactors arranged as shown in the Figure 15.2. The chemical reaction is a

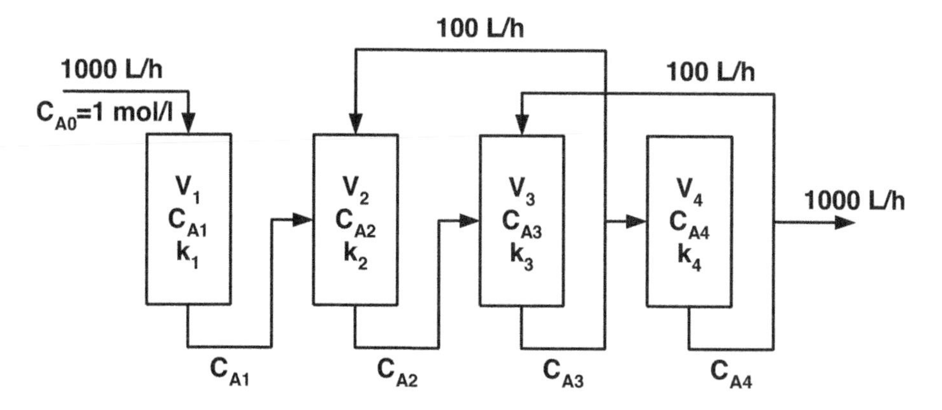


**FIGURE 15.2**
Series of CSTRs

first-order irreversible reaction of the type

$$A \to B. \tag{15.5}$$

The conditions of temperature in each reactor are such that the value of the rate constant $k_1$ is different in each reactor. Also, the volume of each reactor $V_i$ is different. The values of the rate constants and reactor volumes are given in the table below. The following assumptions can be made regarding this system: the system operates at steady state, there are only reactions in the liquid phase, there is no change in volume or density, and the rate of disappearance of component $A$ in each reactor is given by

$$R_i = V_i k_i C_{A,i}. \tag{15.6}$$

Formulate the material balances and solve the resulting system using Gaussian elimination and the Jacobi method. Compare the two methods.

| Reactor | $V_i$ (L) | $k_i$ $(h^{-1})$ | Reactor | $V_i$ (L) | $k_i$ $(h^{-1})$ |
|---|---|---|---|---|---|
| 1 | 1000 | 0.1 | 3 | 100 | 0.4 |
| 2 | 1500 | 0.2 | 4 | 500 | 0.3 |

When we set up the material balances for the above described reactor system we will basically end with a system of 4 coupled linear equations:

$$\begin{bmatrix} 1100 & 0 & 0 & 0 \\ 1000 & -1400 & 100 & 0 \\ 0 & 1100 & -1240 & 100 \\ 0 & 0 & 1100 & -1250 \end{bmatrix} \begin{bmatrix} c_A \\ c_B \\ c_C \\ c_D \end{bmatrix} = \begin{bmatrix} 1000 \\ 0 \\ 0 \\ 0 \end{bmatrix}. \tag{15.7}$$

This system appears to be diagonally dominant and an iterative method such as the Jacobi method could work out here. The following MATLAB code could be used to solve the linear system iteratively:

First we will define the left-hand matrix and the right-hand side:

```
A= [1100 0 0 0; 1000 -1400 100 0; 0 1100 -1240 100; 0 0 1100 -1250];
b= [1000 0 0 0]'
```

Then we will define a diagonal matrix $D$

```
D = diag(diag(A));
```

and a rest matrix $S$:

```
S = A-D;
```

Then we will define a guess solution:

```
xo=[ 0 0 0 0]';
```

We will initialize the error:

```
error = 1;
```

and now we can iterate $x$ with a while-loop until the error decreases below a target value of 0.0001:

```
while error >0.0001
xn = D ^-1*(b-S*xo);
error = norm(xn-xo,2);
xo=xn;
end
```

After 14 iterations the error criterion is met, and the numerical solution to the problem is $c_A = 0.9091$, $c_B = 0.6969$, $c_C = 0.6654$, and $c_D = 0.5855$. In Figure 15.3 the convergence of the solution is plotted with the number of iterations.

## 15.4 PVT behavior of pure substances

$PVT$ behavior of a pure substance is complex and many difficulties arise to describe such behavior by an equation. However, for the gas region, often simple expressions suffice. We could, for example, express PV behavior along an isotherm by means of a power series expansion in $V$. Let's evaluate the following thermodynamic cubic equation of state:

$$PV^3 - (bP + RT)V^2 + aV - ab = 0. \quad (15.8)$$

$a$ and $b$ are constants for a given chemical species. It is important in thermometry to know for which value of $V$ Equation 15.8 goes to zero, for a specified pressure and temperature.

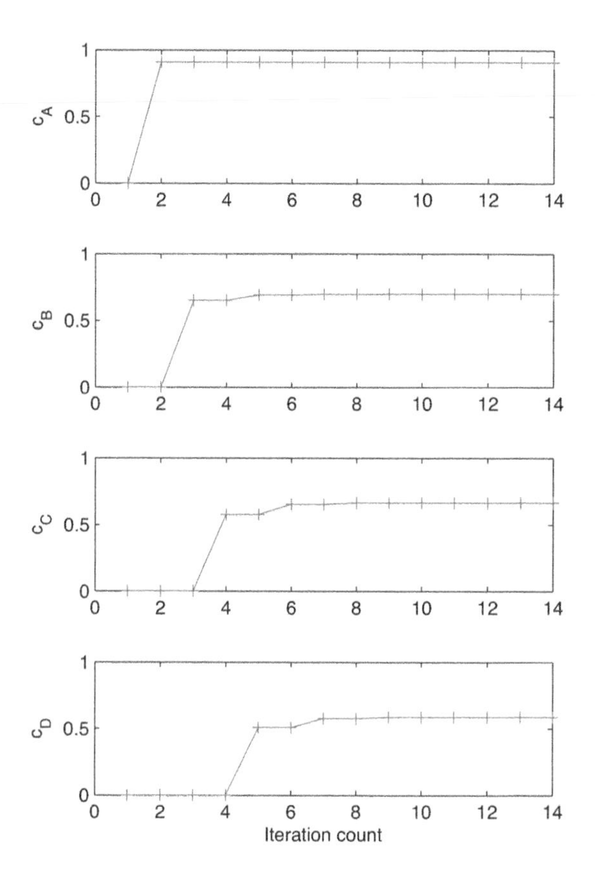


**FIGURE 15.3**
Convergence of the solution with the Jacobi method

We have the following information available:

$$\begin{aligned} a &= 0.0346 \\ b &= 0.0238 \\ R &= 0.08134. \end{aligned}$$

Use Newton's method to determine the $V$'s for which Equation 15.8 goes to zero for the pressure range $P = 1, 2, \cdots 15$ (bar) and temperature range $T = 298, 303$, and 333 (K).

For this assignment we can use the code that was supplied earlier for solving nonlinear equations with Newton's method:

```
function [solution] = Newton(MyFunc,Jacobian,Guess,tol,P,T)
x = Guess;
k =1;
error = 2*tol
while error > tol
 F = feval(MyFunc,x,P,T);
 J = feval(Jacobian,x,P,T);
 dx = J \(-F);
 x= x + dx;
 F = feval(MyFunc,x,P,T);
 error = max(abs(F));
end
solution = x;
return
```

There are slight modifications in the code, as we would like to be able to supply different values for the pressures and temperatures. We now write a MATLAB function that contains the equation of state that we would like to solve:

```
function f = MyFunc(x,P,T)
a = 0.0446;
b = 0.0238;
R = 8.314;
f = P*x. ^3 - (b*P+R*T)*x.^2+ a*x - a*b;
```

Of course we also have to provide an expression for the derivative, which is

$$\left(3PV^2 - 2(bP + RT)V + a\right). \tag{15.9}$$

We will enter this into a MATLAB function as follows:

```
function df = Jacobian(x,P,T)
a = 0.0446;
b = 0.0238;
R = 8.314;
df = 3*P*x.^2 - 2*(b*P+R*T)*x+a;
```

We can now call Newton's method, the equation of state and the derivative of this equation from a main MATLAB script that solves the equation for a series of pressures and temperatures:

```
P = linspace(1e5,15e5,10);
T = linspace(298,333,10);
for i = 1:length(P)
 for j = 1:length(T)
 x(i,j) = Newton(@MyFunc,@Jacobian,0.01,1e-5,P(i),T(j));
 end
end
```

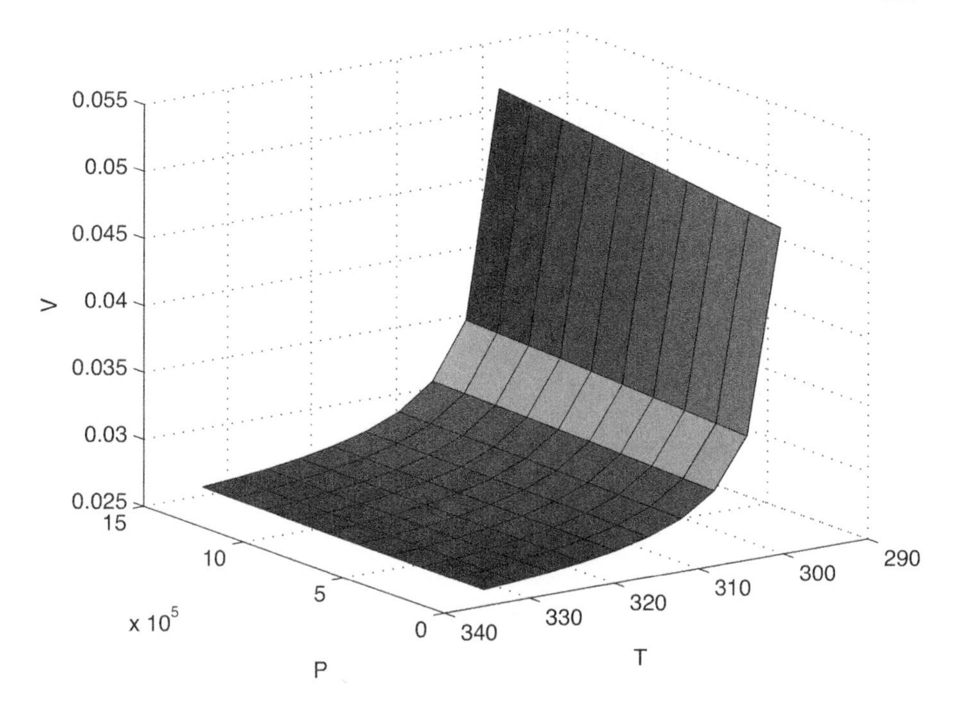


**FIGURE 15.4**
Critical volume as a function of pressure and temperature

The computed volumes can be plotted as a function of pressure and temperature by typing

```
surf(P,T,x);
```

which will give a surface plot as shown in Figure 15.4.

## 15.5 Dynamic modeling of a distillation column

In this assignment we are going to simulate a simple distillation column containing only three stages.

You may assume a binary separation at constant pressure and negligible vapor holdup, the liquid holdup at each stage is 1.0 (kmol), perfect control of levels using $D$ and $B$ ($LV$ configuration), constant molar flows (which replace the energy balance), vapor-liquid equilibrium on all stages, constant relative volatility for the VLE, and constant liquid holdup (i.e., neglect flow dynamics). With these assumptions, the only variables are the mole fraction of $x_i$ of the light component on each stage where $i$ is the stage number.

The column separates a binary mixture with a relative volatility of $\alpha = 10$ and has two theoretical stages ($N = 2$), plus a total condenser. Stage 3 is the total condenser ($D = 0.5 kmol/min$), the liquid feed ($F = 1 kmol/min$) enters on stage 2 with a composition of $z_f = 0.5$, and stage 1 is the reboiler. The reflux ratio is 6.1.

Formulate the material balances and the volatility relations. Given the initial conditions $x_i(0) = 0, \forall i$, use the resulting ODE system to simulate and create composition profiles.

We can start by setting up the component balances, first for the condenser:

$$\frac{d(M_3 x_3)}{dt} = V_2 y_2 - L_3 x_3 - D x_3, \tag{15.10}$$

then for the feed stage:

$$\frac{d(M_2 x_2)}{dt} = x_F F + L_3 x_3 + V_1 y_1 - L_2 x_2 - V_2 y_2, \tag{15.11}$$

and, of course, for the reboiler:

$$\frac{d(M_1 x_1)}{dt} = L_2 x_2 - V_1 y_1 - B x_1. \tag{15.12}$$

We now define $V = V_1$, $L = L_3$, $R = L/D$. We have $D = V - L = L/R$, $V_1 = V_2 = V = (1 + 1/R)L$, $L_2 = L + F$, $B = L + F - V = F - 1/R$. We may assume a constant liquid holdup and a negligible vapor holdup. Of course there is a relationship between the vapor and liquid composition by relative volatilities:

$$y_i = \frac{\alpha x_i}{1 + (\alpha - 1)x_i}. \tag{15.13}$$

In MATLAB we may write a function that contains the above formulated model:

```
function dxdt = distcol(t,x)
R = 6.1
M = 1.0;
D = 0.5;
L = R*D;
F = 1;
xF = 0.5;
alpha = 10;
dxdt = zeros(3,1);
y(1) = (alpha*x(1))/(1+(alpha-1)*x(1));
y(2) = (alpha*x(2))/(1+(alpha-1)*x(2));
y(3) = (alpha*x(3))/(1+(alpha-1)*x(3));
```

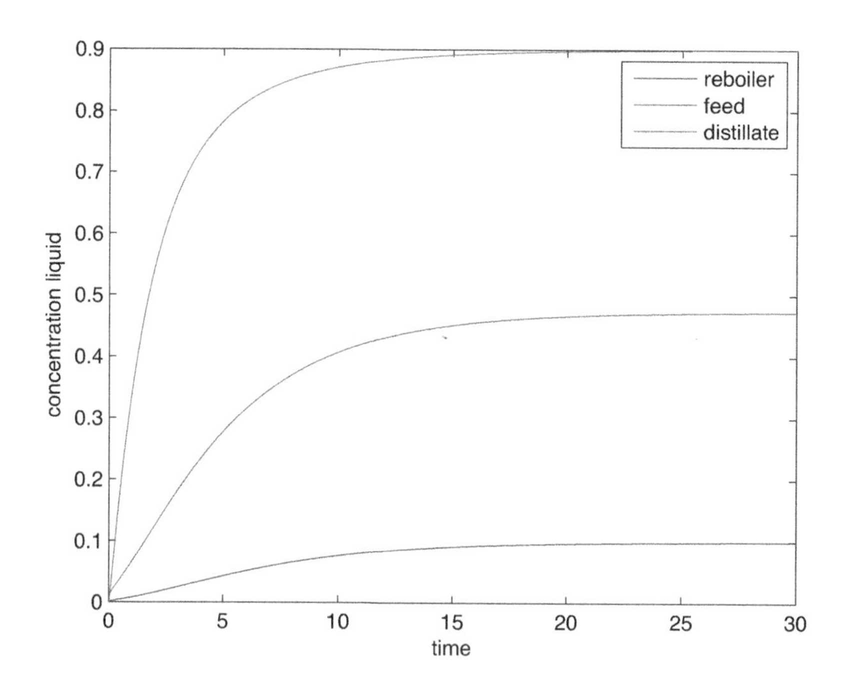


**FIGURE 15.5**
Concentration profiles for the distillation column

```
dxdt(1) = (x(2) + x(1)/R - (1+1/R)*y(1))*L /M + (x(2)-x(1))*F/M;
dxdt(2) = ((x(3)-x(2))+(1+1/R)*(y(1)-y(2)))*L/M+(xF-x(2))*F/M;
dxdt(3) = (1+1^R)*(y(2)-x(3))*L^M;
```

We can now call the ODE solver and solve the system numerically for initial conditions $x_i = 0$, by typing:

```
>>[t,x] = ode45(@distcol,[0 30],[0 0 0])
>>plot(t,x)
```

The resulting concentration profile is given in Figure 15.5.

## 15.6 Dynamic modeling of an extraction cascade (ODEs)

In liquid–liquid extraction contactors, mathematical models must be developed that simulate the dynamic behavior of the equipment [16]. Such a model is important to determine the time required to reach a steady state once a process upset occurs. Consider a three-stage countercurrent contactor (shown

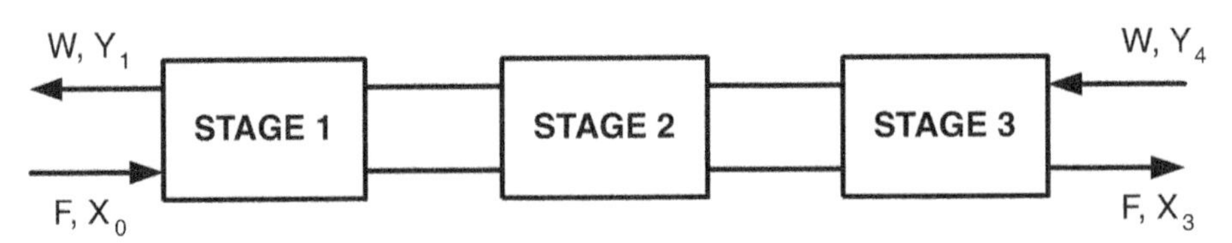


**FIGURE 15.6**
Schematic representation of an extraction stage cascade

in Figure 15.6), where 0.0774 kg/min of kerosene that contains 0.1 kg propanoic acid/kg kerosene is flowing counter-currently to 0.0993 kg/hr water with 0.001 kg propanoic acid/kg water.

At any time, the transfer of solute from the kerosene to the water is described by $Y_n = mX^*_n$, where $Y_n$ is the propanoic acid concentration in the water extract phase (kg propanoic acid/kg water), $X^*_n$ is the equilibrium propanoic acid concentration in the kerosene (kg propanoic acid/kg kerosene), and $m$ is the propanoic acid distribution coefficient =13.4 in this case.

Whether concentration reaches the equilibrium value at each stage is determined by the Murphree efficiency, which is defined as follows:

$$E^0 = \frac{X_{n-1} - X_n}{X_{n-1} - X^*_n}. \tag{15.14}$$

Before the start of the process, the contactors (mixer) contained 0.5 kg of the kerosene phase and 0.3 kg of the water phase. Two kinds of models could be used to model the mass transfer process, which basically involves mass balance on the solute. In an equilibrium-based model, only one equation is sufficient to describe both phases:

$$H_x \frac{dX_n}{dt} + H_y \frac{dY_n}{dt} = F(x_{n-1} - X_n) + W(Y_{n+1} - Y_n). \tag{15.15}$$

$H_x$ is the holdup in the kerosene phase, $Y_n$ is the holdup in the water phase, $F$ is the flow rate of the kerosene raffinate phase, $W$ is the flowrate of the water solvent phase.

In the rate-based model two equations are used to describe the mass balance of the propanoic acid in each phase. For the Raffinate phase

$$H_x \frac{dX_n}{dt} = F(X_{n-1} - X_n) - H_x K_x a \left( X_n - \frac{Y_n}{m} \right) \tag{15.16}$$

and for the extract phase:

$$H_y \frac{dY_n}{dt} = W(Y_{n+1} - Y_n) + H_x K_x a \left( X_n \frac{Y_n}{m} \right). \tag{15.17}$$

(a) Consider a rate-based model and derive a mass balance equation of the solute for each stage and each phase. Plot the concentration profile of the solute in each phase and determine how long it will take before the concentration reaches steady state. Initially the kerosene and the water phase at each stage contains the same concentration of solute as the entering respective phases. Assume an overall mass transfer coefficient of 0.46465 min$^-$1.

(b) Consider an equilibrium-based model and assume a steady-state condition. How does the concentration profile in each stage vary when the Murphree stage efficiency changes from low values close to 0 to high values close to 1? Hint: Derive the steady-state algebraic equations for each stage.

This assignment can be solved in a manner that similar to the distillation example. We start by writing a MATLAB function that contains the balance equations, which are actually given in this assignment:

```
function dcdt=extract(t,c)
Hx = 0.5;
Hy = 0.3;
cxfeed = 0.1;
cyfeed = 0.001;
F = 0.0074;
W = 0.0993;
m = 13.4;
kxa =0.46465;
dcdt=zeros(length(c),1);
k=0.5*length(c)+1;
for i=1:k-1
 if i==1
 dcdt(i,1)=(F/Hx)*(cxfeed-c(i,1))-kxa*(c(i,1)-(c(i+k-1,1)/m));
 else
 dcdt(i,1)=(F/Hx)*(c(i-1,1)-c(i,1))-kxa*(c(i,1)-(c(i+k-1,1)/m));
 end
end
for i=k:length(c)
 if i==length(c)
 dcdt(i,1)=(W/Hy)*(cyfeed-c(i,1))+(Hx/ Hy)*kxa*(c(i-k+1,1)-
 (c(i,1)/m));
 else
 dcdt(i,1)=(W/Hy)*(c(i+1,1)-c(i,1))+(Hx/ Hy)*kxa*(c(i-k+1,1)-
 (c(i,1)/m));
 end
end
```

We can now solve the model with the MATLAB ODE solver for given initial conditions and time span:

```
co=[0.1,0.1,0.1,0.001,0.001,0.001];
tspan=[0:0.05:100];
[t,c]=ode15s(@extract,tspan,co');
```

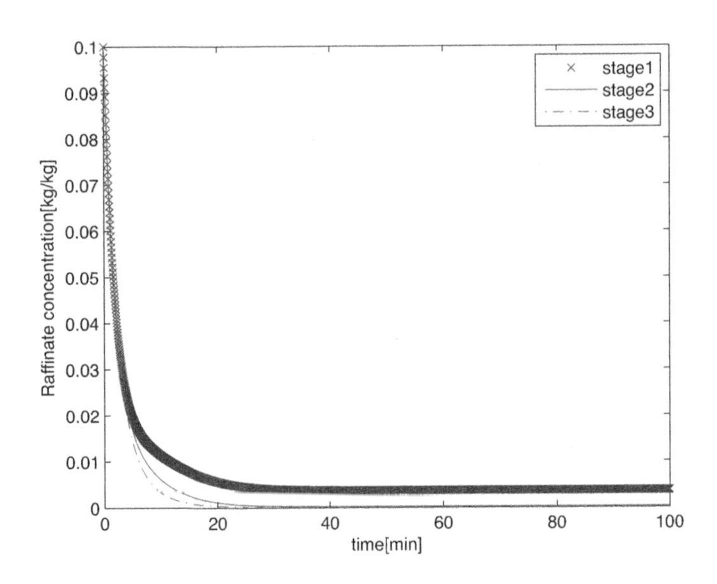


**FIGURE 15.7**
Raffinate concentration with time

In Figures 15.7 and 15.8 the Raffinate concentration and extract concentration as functions of time for different stages are plotted. Now we will solve the equilibrium model (the derivatives are set to zero, resulting in a linear system). In the code below, we first set the values of the model parameters, then create the matrix of the equilibrium model and solve it with the backslash operator.

Set the feed flow rate kg/min

```
F=0.0774;
```

the solvent flow rate kg/min

```
W=0.0993;
```

the concentration solute feed stream

```
Xin=0.1;
```

the concentration solute solvent stream

```
Yin=0.01;
```

the Murphree efficiency

```
E=0.1:0.1:1;
```

the distribution coefficient

```
m=13.5;
```

Now we will initialize the variables

```
x=zeros(3,length(E));
```

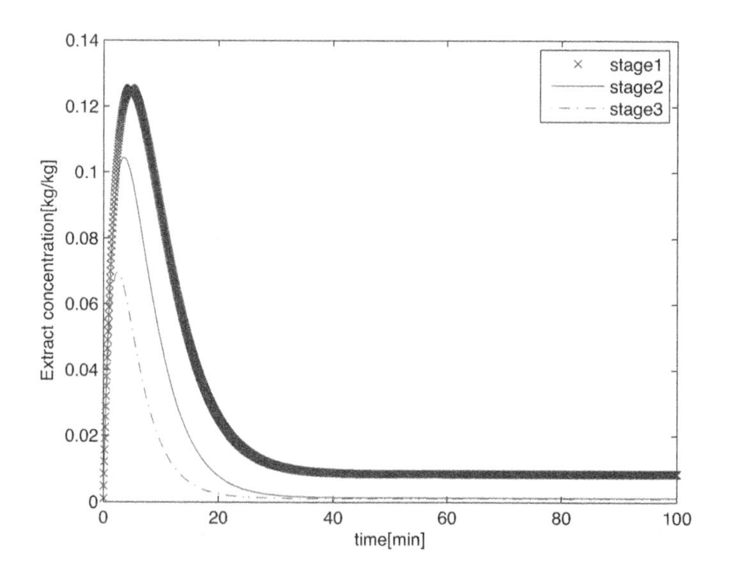


**FIGURE 15.8**
Extract concentration with time

```
y=zeros(3,length(E));
A1=zeros(3,3);
```

We will now define the matrix of the equilibrium model and find the solution with the backslash operation:

```
for i=1:length(E)
A=F-W*m*(1-(1/E(i)));
B=W*m*(1-(1/E(i)))-F-(W*m/E(i));
C=W*m/E(i);
A1=[B C 0; A B C; 0 A (B-(W*m*(1-(1/E(i)))))];
B1=[-A*Xin;0;-W*Yin];
x(:,i)=A1^B1;
for j=1:3
 if j==1
 y(j,i)=m.*Xin*(1-(1/E(i)))+(m/E(i))*x(j,i);
 else
 y(j,i)=m.*x(j-1,i)*(1-(1/E(i)))+(m/E(i))*x(j,i);
 end
 end
end
```

The equilibrium concentrations of the Raffinate and extract are shown in Figures 15.9 and 15.10.

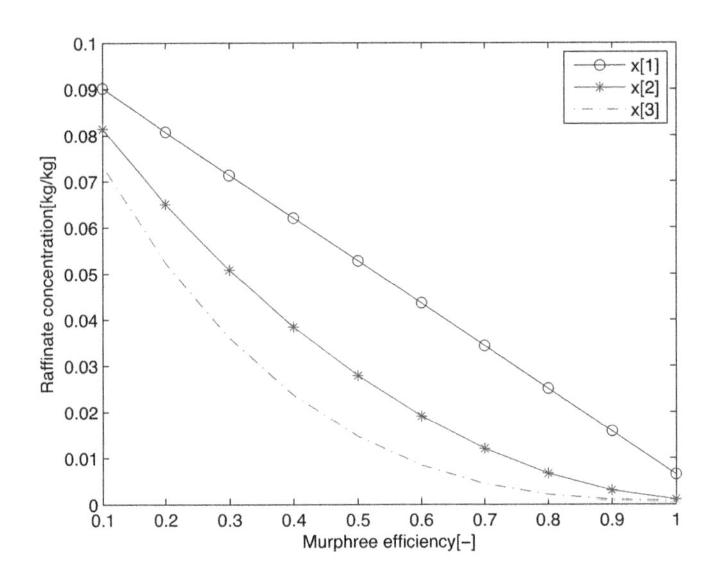


**FIGURE 15.9**
Raffinate concentration with Murphree efficiency

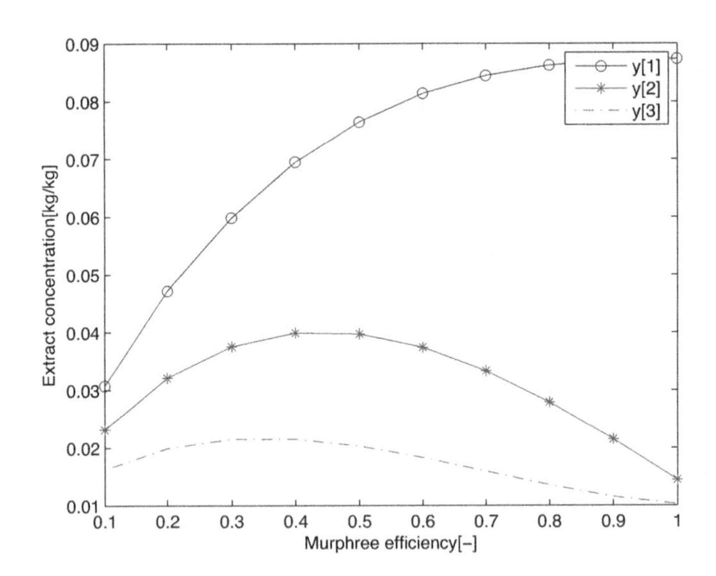


**FIGURE 15.10**
Extract concentration with Murphree efficiency

## 15.7 Distributed parameter models for a tubular reactor

Consider a tubular chemical reactor of length $L$ with two reactions occurring: $A + B \rightarrow C$ and $C + B \rightarrow D$. A fluid with components $A$ and $B$ flows through the reactor with an axial velocity of $v_z$. Besides convection, there is also dispersion taking place in the reactor. The concentration of $A$, $B$, $C$, and $D$ are governed by the coupled set of PDEs

$$\begin{aligned}
\frac{\partial c_A}{\partial t} &= -v_z \frac{\partial c_A}{\partial z} + D\frac{\partial^2 c_A}{\partial z^2} - k_1 c_A c_B \\
\frac{\partial c_B}{\partial t} &= -v_z \frac{\partial c_B}{\partial z} + D\frac{\partial^2 c_B}{\partial z^2} - k_1 c_A c_B - k_2 c_B c_C \\
\frac{\partial c_C}{\partial t} &= -v_z \frac{\partial c_C}{\partial z} + D\frac{\partial^2 c_C}{\partial z^2} + k_1 c_A c_B - k_2 c_B c_C \\
\frac{\partial c_D}{\partial t} &= -v_z \frac{\partial c_D}{\partial z} + D\frac{\partial^2 c_D}{\partial z^2} + k_2 c_B c_C.
\end{aligned}$$

At the reactor inlet and outlet the Danckwerts boundary conditions apply, where the inlet boundary conditions are

$$v_z \left[c_j(0) - c_{j0}\right] - D\left.\frac{dc_j}{dz}\right|_{z=0} = 0,$$

with $j = A, B, C, D$. The outlet boundary conditions are given as:

$$\left.\frac{dc_j}{dz}\right|_{z=L} = 0.$$

Solve this system in MATLAB, using the following data; the length of the tubular reactor $L = 10$, the rate constants, $k_1 = k_2 = 1.0$, the inlet concentrations, $c_{A0} = 1$, $c_{B0} = 1$, $c_{C0} = 0$, $c_{D0} = 0$, the axial velocity, $v_z = 1.0$, the dispersion coefficient $D = 1.10^{-4}$ and the initial conditions $c_A(0) = 2$ and $c_B(0) = c_C(0) = c_D(0) = 0$.

We can solve the tubular reactor PDEs with MATLAB's PDE solver, which is called `pdepe`. It enables the user to solve parabolic-elliptic partial derivatives. In the chapter on PDEs, the use of the pdepe solver was explained. MATLAB needs to receive the model, the boundary conditions, and the initial conditions in a predefined format. For example, the PDE should be defined as:

$$c\frac{\partial u}{\partial t} = x^m \frac{\partial}{\partial x}\left(x^{-m} f\right) + s. \tag{15.18}$$

We should choose $c$, $f$, and $s$ in such a way that it corresponds with the equation you want to solve. In MATLAB we can write a function that contains the tubular reactor model as follows:

```
function [c,f,s] = pdex2pde(z,t,C,dCdz)
k1 = 1.00;
k2 = 1.00;
vz = 1.00;
Dz = 1e-4;
c = [1;1;1;1];
f = [Dz*dCdz(1);
Dz*dCdz(2);
Dz*dCdz(3);
Dz*dCdz(4)];
s = [-vz*dCdz(1)-k1*C(1)*C(2);
-vz*dCdz(2)-k1*C(1)*C(2)-k2*C(2)*C(3);
-vz*dCdz(3)+k1*C(1)*C(2)-k2*C(2)*C(3);
-vz*dCdz(4)+ k2*C(2)*C(3)];
```

In the same fashion, the boundary conditions have to follow a certain formula:

$$p + qf = 0. \tag{15.19}$$

We can now write a MATLAB function that contains the boundary conditions:

```
function [pl,ql,pr,qr] = pdex2bc(zl,Cl,zr,Cr,t)
Cin =[1 1 0 0];
Dz =1e-4;
pl = [Cl(1)-Cin(1);Cl(2)-Cin(2);Cl(3)-Cin(3);Cl(4)-Cin(4)];
ql = [-1;-1;-1;-1];
pr = [0;0;0;0];
qr = [1/Dz;1/Dz;1/Dz;1/Dz];
```

Lastly, we will write a function with the initial conditions:

```
function C0 = pdex2ic(z);
C0=[1 ;1 ;0 ;0];
```

Now we can call the pdepe solver to solve the model, after setting the domain for space and time:

```
>>z = linspace(0,10,30);
>>t = linspace(0,20,30); >>sol = pdepe(0,@pdex2pde,@pdex2ic,@pdex2bc,z,t);
```

In Figure 15.11 the concentration profiles over time and distance are shown and Figure 15.12 shows the steady-state profile.

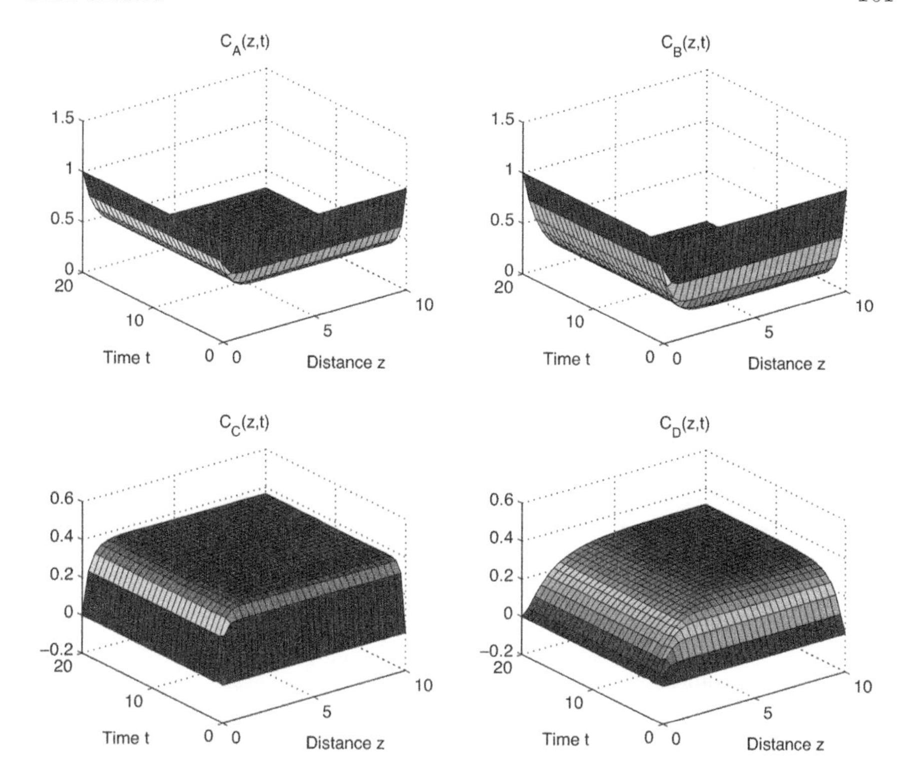


**FIGURE 15.11**
Concentration profiles with time and distance

## 15.8 Modeling of an extraction column

In Figure 15.13, the material balances over a differential length of an L-L extraction contactor are shown, for one-dimensional counter current flow [26]. All relevant data for the extraction column is given in Table 15.1. After derivation of the balances, a set of coupled partial differential equations is obtained, where the component $c_x$ in the light phase may be expressed over the column length $z$ as

$$E_x \frac{\partial^2 c_x}{\partial z^2} - U_x \frac{\partial c_x}{\partial z} - k_{ox} a (c_x - c_x^*) = 0 \tag{15.20}$$

and where the component $c_y$ in the heavy phase can be similarly written as

$$E_y \frac{\partial^2 c_y}{\partial z^2} + U_y \frac{\partial c_y}{\partial z} + k_{ox} a (c_x - c_x^*) = 0. \tag{15.21}$$

The equilibrium relation is given as

$$c_x^* = m c_y. \tag{15.22}$$

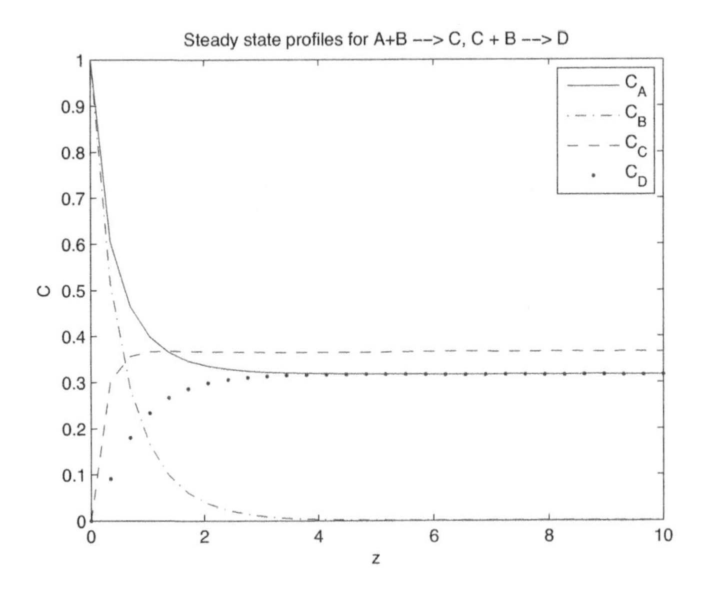


**FIGURE 15.12**
Steady-state profiles

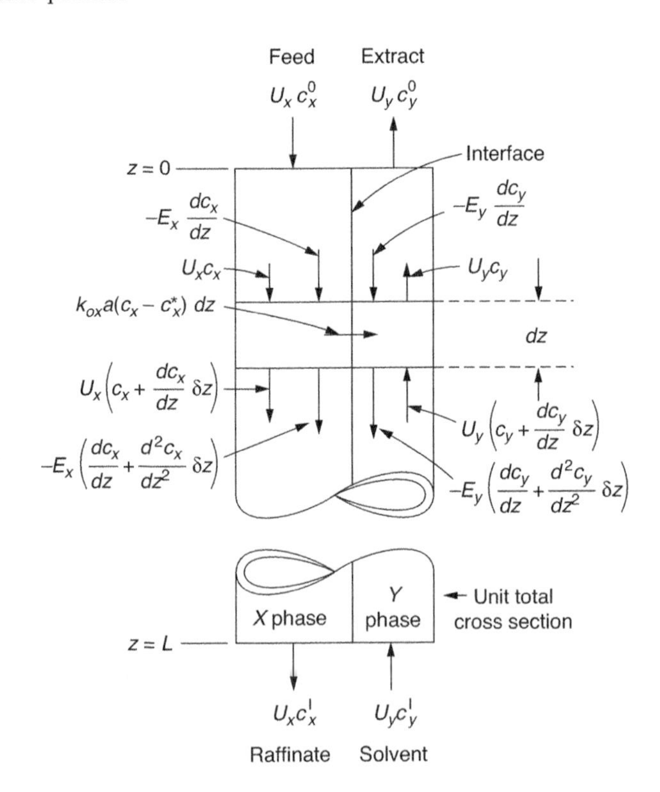


**FIGURE 15.13**
Graphical representation of the extraction column

| | | | |
|---|---|---|---|
| Hold-up x | $U_x$ | $10^{-3}$ | $(m^3/m^2/s)$ |
| Hold-up y | $U_y$ | $10^{-3}$ | $(m^3/m^2/s)$ |
| Axial dispersion x | $E_x$ | $1.8\times10^{-4}$ | $(m^2/s)$ |
| Axial dispersion y | $E_y$ | $1.4\times10^{-4}$ | $(m^2/s)$ |
| Overall mass transfer | $k_{ox}$ | $5.0\times10^{-4}$ | $(m/s)$ |
| Interfacial area | $a$ | 150 | $(m^2/m^3)$ |
| Slope equilibrium line | $m$ | 3.5 | (-) |

**TABLE 15.1**
Extraction column data

For a system in which we are going to extract acrylic acid from water into ethyl acetate by means of a pulsed disk contactor [1], we have the following data available (Table 14.1):

In our model the $z$ coordinate is normalized ($z = l/L$), given the top composition ($c_x(z = 0) = 0.05\ kg/m^3$ and $c_y(z = 0) = 0.05\ kg/m^3$) and the bottom composition ($c_x(z = 1) = 0.2\ kg/m^3$ and $c_y(z = 1) = 1.0\ kg/m^3$). The objective of this assignment is to solve the resulting PDE system as a boundary value problem with the `bvp4c` solver of MATLAB (the bvp solver requires you to rewrite the system as a set of ODEs).

We can convert the PDE system to a set of coupled ODEs. First, let us fill in all known data. The PDE system then turns into

$$C''_x = 5.56C'_x + 416.67C_x - 119.05C_y \quad (15.23)$$
$$C''_y = -7.14C'_y + 535.71C_x + 153.06C_y. \quad (15.24)$$

For reasons of simplicity, let

$$y = [y_1, y_2, y_3, y_4]^T = [C_x, C'_x, C_y, C'_y]^T. \quad (15.25)$$

Now, if we take the derivative of $y$ we get

$$y' = [[y'_1, y'_2, y'_3, y'_4]^T = [C'_x, C''_x, C'_y, C''_y]^T. \quad (15.26)$$

In other words,

$$y'_1 = y_2 \quad (15.27)$$
$$y'_2 = 5.56y_2 + 416.67y_1 - 119.05y_3 \quad (15.28)$$
$$y'_3 = y_4 \quad (15.29)$$
$$y'_4 = -7.14y_4 - 535.71y_1 + 153.06y_3. \quad (15.30)$$

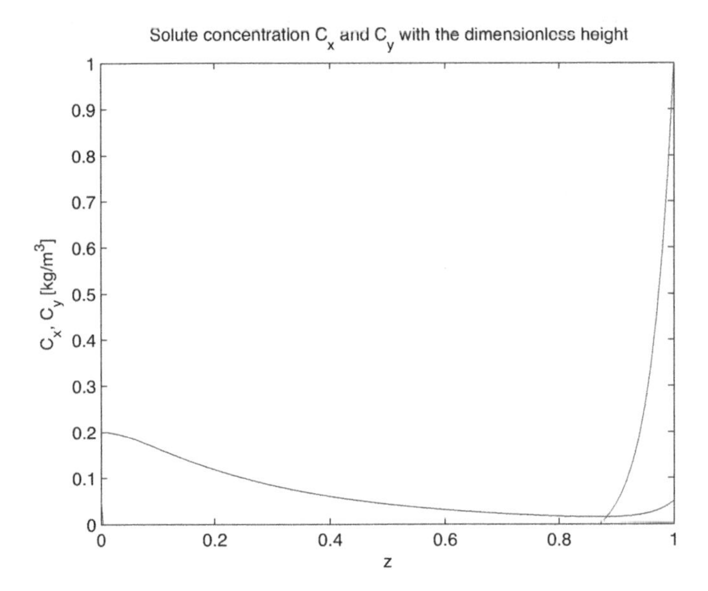


**FIGURE 15.14**
Concentration profile over the column length

We have obtained a system of ODEs with the following boundary conditions $y_a(z = 0) = 0.2$, $y_a(z = L) = 0.05$, and $y_b(z = 0) = 0.05$, and $y_b(z = L) = 1$. We can solve this problem now with the Boundary Value Problem solver of MATLAB, bvp4c.

Similar as with the pdepe solver, we have to write a function with the model and the boundary conditions, which can be done as follows for the model:

```
function F = dEqs(x,y)
F = zeros(1,4);
F(1) = y(2);
F(2) = 5.56*y(2) + 416.67*y(1) - 119.05*y(3);
F(3) = y(4);
F(4) = -7.14*y(4) - 535.71*y(1) + 153.06*y(3);
```

and for the boundary conditions:

```
function r = bvp4bc(ya,yb)
r = [ya(1) - 0.2 ; ya(2) - 0.05; yb(1) - 0.05; yb(2) - 1];
```

The model can now be solved with the boundary value problem solver of MATLAB:

```
>>solinit = bvpinit(linspace(0,1,100),[0.2 0.05 0.05 1]);
>>sol = bvp4c(@dEqs,@bvp4bc,solinit);
```

In Figure 15.14 the concentration profiles over the column length are shown.

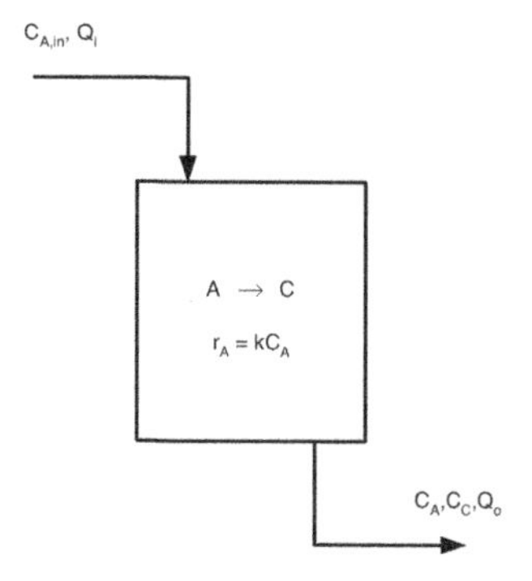


**FIGURE 15.15**
Graphical representation of the reactor

## 15.9 Fitting of kinetic data

In a reactor a reaction takes place, where reactant $A$ is converted into product $C$, according to

$$A \rightarrow C.$$

The reaction rate can be described as

$$r_A = kC_A.$$

Figure 15.15 shows how reactant $A$ enters the reactor with a flow $Q_i$. At the outlet of the reactor, the unreacted $A$ and the formed $C$ leave the reactor with a flow $Q_o = Q_i$. We are able to monitor $A$ and $C$ at the outlet. If we write out the component balances for $A$ and $C$ we obtain the following ODE system:

$$\begin{aligned} \frac{dC_A}{dt} &= \frac{Q_o}{V_r}(c_{A,in} - c_A) - r_A \\ \frac{dC_C}{dt} &= r_A - \frac{Q_o}{V_r}c_C. \end{aligned}$$

Experimental data was collected with $c_A$ and $c_C$ versus time. Further is the following information known $c_{A,in} = 1mol/m^3$ and $Q_o = 1mol/h$.

Determine the kinetic model parameter $k$ and the reactor volume $V_r$ using `lsqnonlin`.

The data is stored in `dataset.zip`.

In Chapter 10 the reader is introduced to `lsqnonlin`. We will first write a MATLAB function that contains the reactor model, which is given above:

```
function dcdt = simpleode(t,c,k);
Q0 = 1;
cAin = 1;
dcdt = zeros(2,1);
dcdt(1) = (Q0/k(1))*(cAin - c(1)) - k(2)*c(1);
dcdt(2) = -(Q0/k(1))*c(2) + k(2)*c(1);
```

We will also have to supply to `lsqnonlin` a fitting criterion function, which we code as the difference between predicted points and experimental data:

```
function error = fitcrit(ke,T,C,Tfinal,C0)
[t,ce] = ode45(@simpleode,T,C0,[],ke);
error = (ce-C).^2;
```

First we will load the data:

```
>>load dataset;
```

We initialize our data:

```
>>C = [cA cC];
>>T = tout;
>>[nrows,ncols] = size(C);
>>Tfinal = T(nrows);
```

We supply the initial values for the concentrations:

```
>>C0 = [0.00 0.00];
```

We also have to come up with initial guesses for the fit parameters:

```
>>k0 = [1.00 1.00];
```

And we can, optionally, also set the lower and upper bounds for model parameters:

```
>>LB = [0.00 0.00];
>>UB = [Inf Inf ];
```

Now we perform nonlinear least squares fit with the `lsqnonlin` routine:

```
>>options = optimset('TolX',1.0E-6,'MaxFunEvals',1000);
>>[ke,RESNORM,RESIDUAL,EXITFLAG,OUTPUT,LAMBDA,JACOBIAN] = ...
lsqnonlin(@fitcrit,k0,LB,UB,options,T,C,Tfinal,C0);
```

The optimized fit parameters are stored in `ke`, which can now be used to simulate the reactor model:

```
>>[te,Ce] = ode45(@simpleode,T,C0,[], ke);
```

The model and the experimental data can now be compared:

```
>>plot(T,C(:,1),'ro',te,Ce(:,1),'r-',T,C(:,2),'b+',te,Ce(:,2),'b-');
xlabel('t');ylabel('c'); legend('data cA','model cA','data cB','model cB');
```

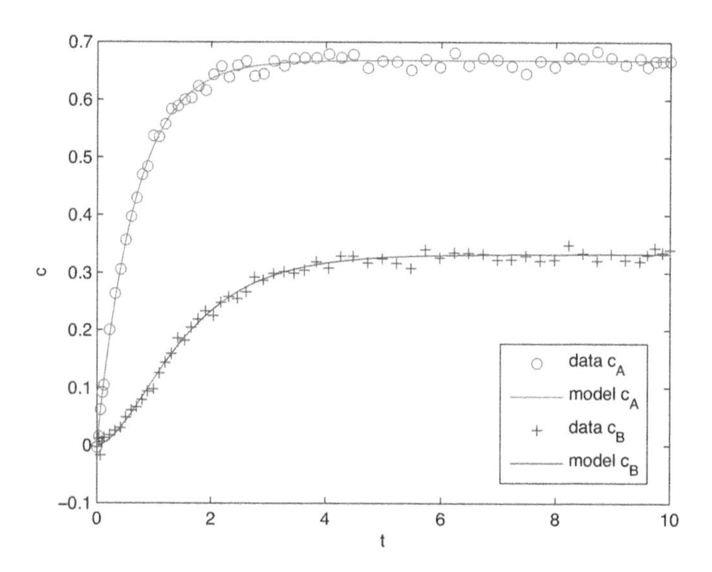


**FIGURE 15.16**
Model and data from regression

Figure 15.16 shows the model and the experimental data. We can also compute the confidence limits of the fitted parameters with `nlparci`:

```
>>cflim=nlparci(ke,RESIDUAL,JACOBIAN);
```

## 15.10 Fitting of NRTL model parameters

Vapor liquid equilibrium data for ethanol (1) + water (2) at 101.3 kPa using a liquid-vapor ebullition-type equilibrium still is reported in Table 15.2. See also [28]. The vapor-liquid equilibrium for every component can be represented by the well-known Raoult's law (Equations 15.31 and 15.32).

$$y_1 P = x_1 \gamma_1 P_1^{sat} \tag{15.31}$$

$$y_2 P = x_2 \gamma_2 P_2^{sat} \tag{15.32}$$

Where $y_i$ and $x_i$ are the molar composition in the vapor and liquid phase for the component $i$, respectively, $P$ represents the total pressure and $P_i^{sat}$ is the vapor pressure. The vapor pressure can be obtained with the Antoine equation (Equation 15.40) with the parameters reported in Table 15.3. Equations 15.31 and 15.32 make use of the activity coefficients ($\gamma_1$ and $\gamma_2$) to describe the

| $x_1$ | $y_1$ | $T/K$ |
|---|---|---|
| 0.056 | 0.366 | 362.19 |
| 0.091 | 0.448 | 359.26 |
| 0.189 | 0.539 | 356.33 |
| 0.286 | 0.582 | 354.89 |
| 0.323 | 0.600 | 354.46 |
| 0.331 | 0.605 | 354.41 |
| 0.419 | 0.627 | 353.59 |
| 0.512 | 0.666 | 352.85 |
| 0.620 | 0.712 | 352.16 |
| 0.704 | 0.759 | 351.74 |
| 0.715 | 0.764 | 351.70 |
| 0.798 | 0.818 | 351.41 |
| 0.843 | 0.851 | 351.37 |
| 0.847 | 0.854 | 351.37 |
| 0.849 | 0.856 | 351.36 |
| 0.884 | 0.886 | 351.34 |
| 0.908 | 0.907 | 351.33 |
| 0.922 | 0.920 | 351.33 |

**TABLE 15.2**
Vapor–liquid equilibrium data for the binary system ethanol (1) + water(2)

| Compound | $A$ | $B$ | $C$ |
|---|---|---|---|
| Ethanol | 7.28781 | 1623.22 | −44.170 |
| Water | 7.19621 | 1730.63 | −39.724 |

**TABLE 15.3**
Antoine parameters

non-ideal behavior and interactions occurring in the liquid phase. Several thermodynamic models have been proposed to correlate and predict the behavior in the liquid phase making use of those activity coefficients. The Non-Random Two Liquid (NRTL) model has been used to predict these properties. This model is shown in Equations 15.33–15.39.

$$ln(\gamma_1) = x_2^2 \left[ \tau_{21} \left( \frac{G_{21}}{x_1 + x_2 G_{21}} \right)^2 + \frac{\tau_{12} G_{12}}{(x_2 + x_1 G_{12})^2} \right] \quad (15.33)$$

$$ln(\gamma_2) = x_1^2 \left[ \tau_{12} \left( \frac{G_{12}}{x_2 + x_1 G_{12}} \right)^2 + \frac{\tau_{21} G_{21}}{(x_1 + x_2 G_{21})^2} \right] \quad (15.34)$$

$$G_{12} = \exp(-\alpha_{12}\tau_{12}) \quad (15.35)$$

$$G_{21} = \exp(-\alpha_{12}\tau_{21}) \quad (15.36)$$

$$\tau_{12} = a_{12} + \frac{b_{12}}{T/K} \tag{15.37}$$

$$\tau_{21} = a_{21} + \frac{b_{21}}{T/K} \tag{15.38}$$

$$\tau_{11} = \tau_{22} = 0 \tag{15.39}$$

As can be seen from the previous equations, the NRTL model applied to binary systems depends on 5 parameters ($a_{12}, b_{12}, a_{21}, b_{21}, \alpha_{12}$). The non-randomness parameter, $\alpha_{12}$, is commonly fixed to:

- $\alpha_{12} = 0.20$, when there is formation of two liquid phases (liquid–liquid equilibrium),
- $\alpha_{12} = 0.30$, for most of the vapor liquid equilibria commonly found in the industry, and
- $\alpha_{12} = 0.48$, for aqueous systems forming vapor liquid equilibrium.

For the system ethanol (1) + water (2), the parameters of the NRTL model that correlate the experimental data reported in Table 15.2 should be regressed in this assignment. First calculate the experimental activity coefficients for each temperature. Then use one of the MATLAB curve-fitting tools to fit the experimental data to the NRTL model. Compare and evaluate the results with

$$\log(P/kPa) = A - \frac{B}{T/K + C}. \tag{15.40}$$

First, we will write a MATLAB function with the NRTL equations for a binary system:

```
function gammai = NRTL(a12,a21,b12,b21,c12,x1,T)
x2 = 1-x1;
t12 = a12 + b12./T;
t21 = a21 + b21./T;
G12 = exp(-c12*t12);
G21 = exp(-c12*t21);
lngamma1 = x2.^2.*(t21.*(G21./(x1+x2.*G21)). ^2...
+t12.*G12./((x2+x1.*G12).^2));
lngamma2 = x1.^2.*(t12.*(G12./(x2+x1.*G12)). ^2...
+t21.*G21./((x1+x2.*G21).^2));
gammai = exp([lngamma1 lngamma2]);
```

Next we want to write a MATLAB function to compute the vapor pressure for ethanol and water:

```
function Pisat = VapPressure(T)
A = [7.28781 7.19621];
B = [1623.22 1730.63];
```

```
C = [-44.170 -39.724];
for i = 1:length(T)
 Pisat(i,:) = 10.^(A-B./(T(i)+C));
end
```

Now we can define a function that contains the fitting criterion, similar to what we did in the previous example:

```
function F = Objective(Param)
global P xi yi T
x1 = xi(:,1);
N = length(T);
a12 = Param(1);
a21 = Param(2);
b12 = Param(3);
b21 = Param(4);
c12 = Param(5);
gammaipred = NRTL(a12,a21,b12,b21,c12,x1,T);
Pisat = VapPressure(T);
yipred = xi.*gammaipred.*Pisat/P;
Fi = (yi-yipred).^2/N;
F = sum(Fi,2);
```

To ensure that we can access all variables, we make them global:

```
>>global P xi yi T
```

We can load our data set:

```
>>load DataVLE.mat
```

The pressure is `>>P = 101.3;`

The ethanol liquid and vapor molar compositions are `>>x1 = Data(:,1);` `>>y1 = Data(:,2);`

The temperature data is

```
>>T = Data(:,3);
```

and the nonrandomness parameter is

```
>>c12 = 0.48;
```

First we calculate the compositions of component 2:

```
>>x2 = 1-x1; y2 = 1-y1;
```

and the formation of the matrices of compositions in liquid and vapor:

```
>>xi = [x1 x2]; yi = [y1 y2];
```

We supply initial guesses for the fit parameters:

```
>>Param0 = [0 0 -2 -2 0.1];
```

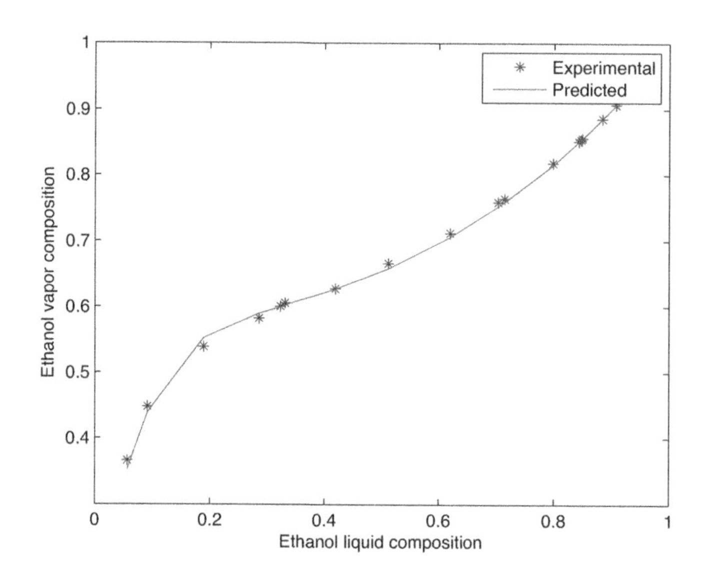


**FIGURE 15.17**
Experimental and predicted compositions

The experimental results and the model are shown in Figure 15.17. We have the possibility to predefine solver settings:

```
>>options = optimset('TolX',1e-8,'MaxIter',1000,'...
MaxFunEvals',2000,'Display','iter','TolFun',1e-8);
```

and now we calculate the fit parameter values with `lsqnonlin`:

```
[Param,feval] = lsqnonlin(@Objective,Param0,[],[],options)
```

We can use the fitted parameters to calculate the predicted activity coefficients:

```
>>gammaipred = NRTL(a12,a21,b12,b21,c12,x1,T);
```

and the vapor pressures:

```
>>Pisat = VapPressure(T);
>>yipred = xi.*gammaipred.*Pisat/P;
```

In Figure 15.16 the predicted and estimated compositions are plotted.

## 15.11 Optimizing a crude oil refinery

A refinery has three crude oils available that have the yields shown in the following table. Because of equipment and storage limitations, production of gasoline (two qualities), kerosene, and fuel oil must be limited as also shown in this table.

There are no plant limitations on the production of other products such as gas oils. The profit on processing crude 1 is \$1.00/*bbl*, on crude 2 it is \$3.50/*bbl*, and on crude 3 it is \$3.0/*bbl*.

Find the approximate optimum daily feed rates of the three crudes to this plant with the simplex method, and compare your outcomes with one of the LP routines available in MATLAB.

| | Vol. percentage yield | | | |
|---|---|---|---|---|
| | Crude 1 | Crude 2 | Crude 3 | max allow. prod. rate |
| Gasoline 1 | 0.30 | 0.31 | 0.20 | 6000 |
| Gasoline 2 | 0.55 | 0.24 | 0.18 | 5500 |
| Kerosene | 0.06 | 0.09 | 0.12 | 2400 |
| Fuel oil | 0.09 | 0.36 | 0.50 | 1200 |

This problem can be formulated as a linear programming problem, which can be solved with the MATLAB `linprog` function. Using the supplied data, the LP is as follows:

$$\begin{aligned}
\min f(x) &= 1.00x_1 + 3.50x_2 + 3.00x_3 \\
s.t. & \\
& 0.30x_1 + 0.31x_2 + 0.20x_3 = 6000 \\
& 0.55x_1 + 0.24x_2 + 0.18x_3 = 5500 \\
& 0.06x_1 + 0.09x_2 + 0.12x_3 = 2400 \\
& 0.09x_1 + 0.36x_2 + 0.50x_3 = 1200.
\end{aligned}$$

The objective represents the profit and the constraints refer to the gasoline types 1 and 2 and the kerosene and fuel oil. This problem can be implemented in MATLAB very easily.

First we define the objective function:

```
>>f = [-1; -3.50; -3];
```

Subsequently we define the constraints: `>>A = [0.3 0.31 0.20;0.55 0.24 0.18;0.06 0.09 0.12;0.09 0.36 0.5];`

```
>>b = [6000; 5500; 2400; 1200];
```

We define the lower bounds:

```
lb = zeros(3,1);
```

and ultimately we call `linprog`:

```
>>[x,fval,exitflag,output,lambda]=linprog(f,A,b,[],[],lb)
```

From this we find that the profit is 12866 USD where $x_1 = 9591.8$, $x_2 = 935.4$, and $x_3 = 0.00$.

## 15.12 Planning in a manufacturing line

From [18] the following problem was taken: A manufacturing line makes two products. Production and demand data are shown in Tables 15.4 and 15.5. Total time available (for production and setup) in each week is 80 h. Starting inventory is zero, and inventory at the end of week 4 must be zero. Only one product can be produced in any week, and the line must be shut down and cleaned at the end of each week. Hence the setup time and costs are incurred for a product in any week in which that product is made. No production can take place while the line is being set up. Formulate and solve this problem as maximizing total net profit over all products and periods.

Although MATLAB has a tool for solving integer optimization problems (`bintprog`), this solver is not very powerful. For this reason, we are going to use another software to formulate and solve the above introduced mixed integer program. First we will formulate the objective:

$$\max Z = \sum_k \sum_t \left( SP_k xs_{k,t} - SC_k u_{k,t} - PC_k xp_{k,t} - IC_k xi_{k,t} - FC_k xf_{k,t} \right). \tag{15.41}$$

| | Product 1 | Product 2 |
|---|---|---|
| Set-up time (h) | 6.00 | 11.00 |
| Set-up costs ($) | 250.00 | 400.00 |
| Production time/unit (h) | 0.50 | 0.75 |
| Production cost/unit ($) | 9.00 | 14.00 |
| Inventory holding cost/unit | 3.00 | 3.00 |
| Penalty cost for unsatisfied demand/unit ($) | 15.00 | 20.00 |
| Selling price ($/unit) | 25.00 | 35.00 |

**TABLE 15.4**
Production data

| Product | Week 1 | Week 2 | Week 3 | Week 4 |
|---|---|---|---|---|
| 1 | 75 | 95 | 60 | 90 |
| 2 | 20 | 30 | 45 | 30 |

**TABLE 15.5**
Demand data

This is a profit function in which $SP_k$ is the sales price of product $k$, $xs_{k,t}$ is the number of sold units of $k$ in week $t$, $SC_k$ are the setup costs, $PC_k$ are the production costs, $IC_k$ are the inventory holding costs, and $FC_k$ is a penalty fee for unsatisfied demand. The variable $u_{k,t}$ is the selector for a product $k$ at time $t$, which is binary. Then there is a variable $xp_{k,t}$, which is the number of produced units, and $xi_{k,t}$, which is the number of units in inventory and $xf_{k,t}$ is the number of units for the unsatisfied demand.

We now introduce a constraint that ensures that there can only be one product produced at any time:

$$\sum_k u_{k,t} = 1. \tag{15.42}$$

We have a production time constraint:

$$PT_k xp_{k,t} \leq (80 - ST_k) u_{k,t}, \tag{15.43}$$

where $PT_k$ is the production time and $ST_k$ is the setup time. Then we keep an inventory balance of

$$xi_{k,t} = xp_{k,t} - xs_{k,t} + xi_{k,t-1}. \tag{15.44}$$

We further know that the inventory has to be zero at the end of week 4:

$$xi_{k,t=wk4} = 0. \tag{15.45}$$

We keep a penalty balance of

$$xf_{k,t} = D_{k,t} - xs_{k,t}, \tag{15.46}$$

where $D_{k,t}$ is the demand, which should match the number of sold units:

$$xs_{k,t} \leq D_{k,t}. \tag{15.47}$$

A typical listing of a GAMS code is shown in Figure 15.18. From GAMS follows a production plan as follows:

| | WK1 | WK2 | WK3 | WK4 |
|---|---|---|---|---|
| P1 | 100 | 100 | — | 90 |
| P2 | — | — | 75 | — |

This production plan leads to a profit of $3360 USD.

```
Sets
        i products /Prod1*Prod2/
        t weeks /wk1*wk4/;
Parameter
        ST(i) Set-up time (hrs)
               /Prod1  6.00
                Prod2 11.00/;
Parameter
        SC(i) Set-up costs ($)
               /Prod1  250.00
                Prod2  400.00/;
Parameter
        PT(i) Production time per unit (hrs)
               /Prod1  0.50
                Prod2  0.75/;
Parameter
        PC(i) Production costs per unit ($)
               /Prod1  9.00
                Prod2 14.00/;
Parameter
        IC(i) Inventory holding  costs for unsatisfied demand per unit ($)
               /Prod1  3.00
                Prod2  3.00/;
Parameter
        PenCost(i) Penalyt costs for unsatisfied demand per unit ($)
               /Prod1  15.00
                Prod2  20.00/;
Parameter
        SP(i) Selling price per unit ($)
               /Prod1  25.00
                Prod2  35.00/;
Table Demand(i,t) Demand data
              wk1 wk2 wk3 wk4
        Prod1 75  95  60  90
        Prod2 20  30  45  30;
Variables
        Z         net profit
        x(i,t)    quantitiy of product i produced in week t
        u(i,t)    Binary variables
        Inventory(i,t)   Inventory;
binary variable u;
Positive variables x,Inventory;
Equations
        Profit        net profit
        Inventorycon(i,t) inventory constraint
        StartInventory(i,t) Inventory at start
        EndInventory(i,t) Inventory at the end
        Timecon(t)       time constraint;
        Profit .. Z =e= sum((i,t),u(i,t)*x(i,t)*SP(i)-u(i,t)*SC(i)-u(i,t)*PC(i)*x(i,t)
        -IC(i)*Inventory(i,t)-(1-u(i,t))*0*PenCost(i)*(Demand(i,t)-x(i,t)));
        Inventorycon(i,t)  .. Inventory(i,t) =e= Inventory(i,t-1) + x(i,t) - Demand(i,t);
        Timecon(t)         ..  sum((i), PT(i)*x(i,t)+ST(i))=l=80;
        StartInventory(i,t).. Inventory(i,'wk1') =e= 0;
        EndInventory(i,t) ..  Inventory(i,'wk4') =e= 0;
model Manufacture /all/;
option MINLP= DICOPT;
solve Manufacture using MINLP maximizing Z;
display x.l, x.m, u.l, u.m;
```

**FIGURE 15.18**
GAMS listing of the Mixed Integer Program

# Bibliography

[1] Asbjornsen, O. A. and T. Hertzberg (1974). *"Constrained regression in chemical engineering practice."* Chemical Engineering Science **29(3)**: 679.

[2] Ataie-Ashtiani, B. and S. A. Hosseini (2005). *"Error analysis of finite difference methods for two-dimensional advection-dispersion-reaction equation."* Advances in Water Resources **28(8)**: 793.

[3] Baleo, J. N. and P. Le Cloirec (2000). *"Validating a prediction method of mean residence time spatial distributions."* AIChE Journal **46(4)**: 675.

[4] Balendra, S. and I. D. L. Bogle (2009). *"Modular global optimisation in chemical engineering."* Journal of Global Optimization **45(1)**: 169.

[5] Biegler, L. T. and I. E. Grossmann (2004). *"Retrospective on optimization."* Computers and Chemical Engineering **28(8)**: 1169.

[6] Biegler, L. T., I. E. Grossmann, et al. (1997). *Systematic Methods of Chemical Process Design*, Prentice Hall.

[7] Botte, G. G., J. A. Ritter, et al. (2000). *"Comparison of finite difference and control volume methods for solving differential equations."* Computers and Chemical Engineering **24(12)**: 2633.

[8] Burghoff, B., E. Zondervan, et al. (2009). *"Phenol extraction with Cyanex 923: Kinetics of the solvent impregnated resin application."* Reactive and Functional Polymers **69(4)**: 264.

[9] Carmo Coimbra, M. D., C. Sereno, et al. (2000). *"Modelling multicomponent adsorption process by a moving finite element method."* Journal of Computational and Applied Mathematics **115(1–2)**: 169.

[10] Caussignac, P. and R. Touzan (1990). *"Solution of three-dimensional boundary layer equations by a discontinuous finite element method, part I: Numerical analysis of a linear model problem."* Computer Methods in Applied Mechanics and Engineering **78(3)**: 249.

[11] Chen, H. S. and M. A. Stadtherr (1984). *"On solving large sparse nonlinear equation systems."* Computers and Chemical Engineering **8(1)**: 1.

[12] Chen, W. S., B. R. Bakshi, et al. (2004). *"Bayesian estimation via sequential Monte Carlo sampling: Unconstrained nonlinear dynamic systems."* Industrial and Engineering Chemistry Research **43(14)**: 4012.

[13] Coimbra, M. D. C., C. Sereno, et al. (2004). *"Moving finite element method: Applications to science and engineering problems."* Computers and Chemical Engineering **28(5)**: 597.

[14] Cruz, P., J. C. Santos, et al. (2005). *"Simulation of separation processes using finite volume method."* Computers and Chemical Engineering **30(1)**: 83.

[15] de Jong, M. C., R. Feijt, et al. (2009). *"Reaction kinetics of the esterification of myristic acid with isopropanol and n-propanol using p-toluene sulphonic acid as catalyst."* Applied Catalysis A: General **365(1)**: 141.

[16] Des Costello (1992). Dynamic Modelling of a Small Scale Liquid-Liquid Extraction Rig, Ph.D. thesis, Edinburgh University.

[17] Dormand, J. R. and P. J. Prince (1980). *"A family of embedded Runge-Kutta formulae."* Journal of Computational and Applied Mathematics **6(1)**: 19–26.

[18] Edgar, T. F., D. M. Himmelblau, et al. (2001). *Optimization of Chemical Processes*, McGraw-Hill.

[19] Evangelista, F. (2005). *"Dynamics of shell and tube heat exchangers: New insights and time-domain solutions."* 7th World Congress of Chemical Engineering, GLASGOW2005, incorporating the 5th European Congress of Chemical Engineering.

[20] Eykhoff, P. (1974). *System Identification: Parameter and State Estimation*, Wiley & Sons.

[21] Floudas, C. A. and X. Lin (2004). *"Continuous-time versus discrete-time approaches for scheduling of chemical processes: A review."* Computers and Chemical Engineering **28(11)**: 2109.

[22] Golub, G. H. and C. F. v. Loan (1996). *Matrix Computations*, The John Hopkins University Press.

[23] Gritton, K. S., J. D. Seader, et al. (2001). *"Global homotopy continuation procedures for seeking all roots of a nonlinear equation."* Computers and Chemical Engineering **25(7–8)**: 1003.

[24] Hahn, B.H. and Valentine, D.T. (2010). *Essential MATLAB for Engineers and Scientists*, Academic Press.

[25] Hangos, K. and Cameron, I. (2001). *Process Modeling and Model Analysis*, Academic Press.

[26] Hartland, S. and A. Kumar (1997). *Design of Liquid-Liquid Extractors*, Mineral Processing and Extractive Metallurgy Review, **17(1)**, 43–79

[27] Ierapetritou, M. G. and C. A. Floudas (1998). *"Effective continuous-time formulation for short-term scheduling. 1. Multipurpose batch processes."* Industrial and Engineering Chemistry Research **37(11)**: 4341.

[28] Kiyofumi Kurihara, Mikiyoshi Nakamichi, and Kazuo Kojima (1993). *Isobaric Vapor-Liquid Equilibria for Methanol + Ethanol + Water and the Three Constituent Binary Systems*, J. Chem. Eng. Data **38**, 446–449.

[29] Koster, L. G., E. Gazi, et al. (1992). *"Finite elements for near-singular systems - an overview."* Computers & Chemical Engineering **16(Supplement 1)**: S43.

[30] Kruger, U., Y. Zhou, et al. (2008). *"Robust partial least squares regression: Part I, algorithmic developments."* Journal of Chemometrics **22(1)**: 1.

[31] Kumar, J., G. Warnecke, et al. (2009). *"Comparison of numerical methods for solving population balance equations incorporating aggregation and breakage."* Powder Technology **189(2)**: 218.

[32] Kuno, M. and J. D. Seader (1988). *"Computing all real solutions to systems of nonlinear equations with a global fixed-point homotopy."* Industrial and Engineering Chemistry Research **27(7)**: 1320.

[33] Liu, F. and S. K. Bhatia (2001). *"Solution techniques for transport problems involving steep concentration gradients: Application to noncatalytic fluid solid reactions."* Computers and Chemical Engineering **25(9–10)**: 1159.

[34] Lohmann, T., H. G. Bock, et al. (1992). *"Numerical methods for parameter estimation and optimal experiment design in chemical reaction systems."* Industrial & Engineering Chemistry Research **31(1)**: 54.

[35] Martinez, E. C., M. D. Cristaldi, et al. (2009). *"Design of dynamic experiments in modeling for optimization of batch processes."* Industrial and Engineering Chemistry Research **48(7)**: 3453.

[36] Marwuardt, D. W. (1963). *"An algorithm for least-squares estimation of nonlinear parameters."* J. Soc. Indust. Appl. Math. **11(2)**: 431-441.

[37] McCann, N. and M. Maeder (2009). *"Tutorial: The modelling of chemical processes."* Analytica Chimica Acta **647(1)**: 31.

[38] Mosbach, S. and A. G. Turner (2009). *"A quantitative probabilistic investigation into the accumulation of rounding errors in numerical ODE solution."* Computers & Mathematics with Applications **57(7)**: 1157.

[39] Motz, S., A. Mitrovic, et al. (2002). *"Comparison of numerical methods for the simulation of dispersed phase systems."* Chemical Engineering Science **57(20)**: 4329.

[40] Murthi, M., L. D. Shea, et al. (2009). *"Numerical problems and agent-based models: For a mass transfer course."* Chemical Engineering Education **43(2)**: 153.

[41] Oberkampf, W. L., S. M. DeLand, et al. (2002). *"Error and uncertainty in modeling and simulation."* Reliability Engineering & System Safety **75(3)**: 333.

[42] Patankar, S. V. and D. B. Spalding (1972). *"A calculation procedure for heat, mass and momentum transfer in three-dimensional parabolic flows."* International Journal of Heat and Mass Transfer **15(10)**: 1787.

[43] Pistikopoulos, E. N. and M. G. Ierapetritou (1995). *"Novel approach for optimal process design under uncertainty."* Computers and Chemical Engineering **19(10)**: 1089.

[44] Sahinidis, N. V. (2004). *"Optimization under uncertainty: State-of-the-art and opportunities."* Computers and Chemical Engineering **28(6–7)**: 971.

[45] Schittkowski, K. (2004). *"Data fitting in partial differential algebraic equations: Some academic and industrial applications."* Journal of Computational and Applied Mathematics **163(1)**: 29.

[46] Schuermans, M., I. Markovsky, et al. (2007). *"An adapted version of the element-wise weighted total least squares method for applications in chemometrics."* Chemometrics and Intelligent Laboratory Systems **85(1)**: 40.

[47] Seth, D., A. Sarkar, et al. (2005). *"Uncertainties in the simulation of catalytic distillation process: A systematic grid refinement study."* Chemical Engineering Science **60(20)**: 5445.

[48] Skogestad, S. (2004). *"Dynamics and control of distillation columns: A tutorial introduction."* Trans. IChemE **75(PART A)**: 539.

[49] Sousa Jr., R., D. M. dos Anjos, et al. (2008). *"Modeling and simulation of the anode in direct ethanol fuels cells."* Journal of Power Sources **180(1)**: 283.

[50] Zitney, S. E. and M. A. Stadtherr (1993). *"Frontal algorithms for equation-based chemical process flowsheeting on vector and parallel computers."* Computers and Chemical Engineering **17(4)**: 319.

[51] Zondervan, E., B. H. L. Betlem, et al. (2007). *"Development of a dynamic model for cleaning ultra filtration membranes fouled by surface water."* Journal of Membrane Science **289(1–2)**: 26.

# Index

**P**